# Practical Arduino Robotics

A hands-on guide to bringing your robotics ideas to life using Arduino

**Lukas Kaul**

BIRMINGHAM—MUMBAI

# Practical Arduino Robotics



**Group Product Manager**: Rahul Nair
**Publishing Product Manager**: Surbhi Suman
**Senior Editor**: Arun Nadar
**Technical Editor**: Arjun Varma
**Copy Editor**: Safis Editing
**Project Coordinator**: Ashwin Kharwa
**Proofreader**: Safis Editing
**Indexer**: Tejal Daruwale Soni
**Production Designer**: Shankar Kalbhor
**Marketing Coordinator**: Nimisha Dua

First published: March 2023

Production reference: 1170223

Published by Packt Publishing Ltd.
Livery Place
35 Livery Street
Birmingham
B3 2PB, UK.

ISBN 978-1-80461-317-7

www.packtpub.com

*To all teachers and mentors, who educate and inspire current and future generations to follow their passions and develop their talents.*

*To my wife, Alice, and my son, Leon, who inspire and teach me every day about the most important things in life.*

# Contributors

## About the author

**Lukas Kaul** is a robotics research scientist, currently working at the Toyota Research Institute in Silicon Valley, where he develops mobile manipulation technologies to support people in their homes and in their workplaces. Throughout his career, he has worked on projects as diverse as humanoid robots, aerial robots, and mobile manipulation systems. A maker at heart, Lukas has been using Arduino technology extensively for more than a decade in countless side-projects as well as in his professional work, for tasks as diverse as building robots that can map underground caves, automating component tests, and creating redundant safety systems. Lukas is passionate about teaching robotics with Arduino to inspire and empower anyone who wants to enter the exciting field of robotics. He holds a Ph.D. from the **Karlsruhe Institute of Technology** (**KIT**), Germany.

# About the reviewers

**Anmol Sarin** is a product designer by profession and is involved with research in robotics and industrial engineering. He has a background in mechanical engineering with a focus on design and manufacturing innovation. His research interests include humanoid robots, artificial muscles, and ASRS systems used in the warehouse industry. Professionally, he has successfully led the full-cycle product design and development of a few award-winning, featured, and patented products in the plumbing fixture and HVAC industry. He shares his unique experience through writing and speaking engagement via different platforms and community associations.

**Pascal P. R. Weiner** received his M.Sc. degree in computer science with distinction from the **Karlsruhe Institute of Technology** (**KIT**), Karlsruhe, Germany, in 2016. He is currently a research scientist at the Institute for Anthropomatics and Robotics at the KIT and a member of the High Performance Humanoid Technologies Lab. His research interests are in intelligent mechanics, electrical systems, and software for grasping in prosthetic and robotic applications. He serves as one of the lead engineers for the design of multiple humanoid robots and instructs practical courses on the electrical design and programming of microcontroller-based mobile robots.

# Table of Contents

# 3

## Making Your Robot Move and Interact with the World with Actuators 55



# 4

## Selecting the Right Arduino Board for Your Project 75

# Part 2: Writing Effective and Reliable Robot Programs for Arduino

5

## Getting Started with Robot Programming 95



6

## Understanding Object-Oriented Programming and Creating Arduino Libraries 121

## 7

## Testing and Debugging with the Arduino IDE 135



# Part 3: Building the Hardware, Electronics, and UI of Your Robot

## 8

## Exploring Mechanical Design and the 3D Printing Toolchain 155

9

## Designing the Power System of Your Robot 173



10

## Working with Displays, LEDs, and Sound 189



11

## Adding Wireless Interfaces to Your Robot 209

# Part 4: Advanced Example Projects to Put Your Robotic Skills into Action

# 13



# 14

# Preface

Robots capture our imagination in stories and in the real world. Many of us are fascinated by these machines that can act autonomously with superhuman strength, speed, or other impressive qualities. Robots are also incredibly useful in many applications, from manufacturing to entertainment and education. Creating complete robot systems is a feat of engineering, and this book strives to enable you to do just that. With the most recent advances in electronics and the availability of more and more open source software tools for robotics, building robots has become more accessible than ever before. Using the popular, open source Arduino platform and easily accessible hardware components, this book will teach you all that is needed to bring your robot ideas to life.

DIY robotics with Arduino is a popular hobby for makers and electronics hobbyists, and without a solid foundation, all of the content you can find on the internet can be overwhelming. This book is based on more than a decade of experience of using Arduino for robotics projects and teaching the underlying technologies. It condenses many hard-earned lessons, tips, and best practices to accelerate your journey to becoming a well-rounded robot creator. With many example projects, from blinking an LED to a radio-controlled, self-balancing telepresence robot, it gives you invaluable hands-on experience that you need for planning your future projects.

Robotics and automation is a booming industry, even beyond its traditional applications. As societies around the world are aging, robotics technology will play a critical role in maintaining prosperity and happiness. With this book, you can lay the foundation for your involvement in robotics and set yourself on the path towards helping to shape our future.

## Who this book is for

Whether you want to build your own robots to solve a specific problem; use robotics projects as a way to learn more about electrical, mechanical, or computer engineering to advance your career; or teach robotics with hands-on examples, you will find this book highly useful.

If you fit into one of the following groups, this book is definitely for you:

- Motivated high school students (and their teachers) that are fascinated by robots and want to learn about robotics hands-on
- Engineering students who want to look beyond their immediate field and see how different engineering disciplines come together to create fascinating systems
- Hobbyists, enthusiasts, and makers that want to learn how to harness the power of Arduino in their own robotics projects, from automating the landing gear of radio-controlled airplanes to DIY home automation

- Professionals who can use Arduino robotic technologies to improve their workflows, enhance productivity, and spark their creativity by learning new skills

## What this book covers

*Chapter 1*, *Introducing Robotics and the Arduino Ecosystem*, gets us on the same page about what we mean by robotics, explains in detail what an Arduino board is, and walks you through setting up the Arduino IDE. By the end of this chapter, your first Arduino program will be running on your Arduino Uno.

*Chapter 2*, *Making Robots Perceive the World with Sensors*, introduces various sensors that are commonly used in DIY robotics and the electrical interfaces they use to provide their data to the Arduino board. The chapter includes detailed code examples that demonstrate their use.

*Chapter 3*, *Making Your Robot Move and Interact with the World with Actuators*, gives you a broad overview of the different electrical actuators that you can choose from for your robotics projects. Example setups and code show you how to use them with Arduino.

*Chapter 4*, *Selecting the Right Arduino Board for Your Project*, packs in a lot of information about the distinct characteristics of Arduino boards and teaches you all you need to know to choose the right board for your project.

*Chapter 5*, *Getting Started with Robot Programming*, introduces important concepts and best practices specific to robot programming that make your code performant, useful, and reusable. Using these concepts from the beginning will save you a lot of time in debugging and troubleshooting, and you will instead develop high-quality robot programs.

*Chapter 6*, *Understanding Object-Oriented Programming and Creating Arduino Libraries*, teaches you how to use one of the most powerful features of the C++ language: object orientation. Using this new skill, the chapter walks you through the process of turning your Arduino program into a cleanly packaged library so that you can effortlessly reuse its functionalities in future projects without having to duplicate any code.

*Chapter 7*, *Testing and Debugging with the Arduino IDE*, is a detailed user guide to the Arduino IDE's most powerful tools for debugging and creating interactive programs, the Serial Monitor, and the Serial Plotter. It also sums up important lessons and best practices that help you avoid common pitfalls and streamline your development process.

*Chapter 8*, *Exploring Mechanical Design and the 3D Printing Toolchain*, switches gears from software to hardware. This chapter introduces you to various techniques that you can use to build your physical robots, with a special emphasis on 3D printing.

*Chapter 9*, *Designing the Power System of Your Robot*, is our excursion into the world of power electronics. It will introduce you to the key considerations when it comes to designing a power system that is the backbone of your robot's safe and reliable operation.

*Chapter 10, Working with Displays, LEDs, and Sound*, helps you make your robots truly interactive. Learning how to change the brightness and color of lights, make a variety of sounds, and display text on an LCD screen allows you to add a lot of personality to your robots.

*Chapter 11, Adding Wireless Interfaces to Your Robot*, introduces you to different types of wireless interfaces that are suitable for and easy to use in DIY robotics projects, including Bluetooth and Wi-Fi. It teaches you about their specific pros and cons so you can decide which interface is best for your own projects.

*Chapter 12, Building an Advanced Line Following Robot Using a Camera*, is a complete case study that puts a lot of what we have learned in the previous chapters to use. At the end of this chapter, you will have a fully functioning, sturdy, and capable mobile robot with a camera and a Bluetooth interface that you can use for many more experiments beyond line following.

*Chapter 13, Building a Self-Balancing, Radio-Controlled Telepresence Robot*, is another case study that takes the line follower and turns it into a self-balancing robot. It highlights additional aspects of robot programming, with an emphasis on feedback control and systematic parameter tuning.

*Chapter 14, Wrapping Up, Next Steps, and a Look Ahead*, recaps what we learned in the previous chapters and puts it in the larger context of the vast field of robotics. It illustrates different options to keep learning about robotics, helps you choose your focus, and offers advice for those who wish to make robotics their career.

## To get the most out of this book

You will need basic familiarity with electrical circuits and Ohm's law, and you should have some experience with basic concepts of computer programming (variables, functions, and loops). Even if you do not have this knowledge, you can still read this book and learn the required skills on the fly. Each chapter has pointers to further reading that you can use to deepen your understanding or as starting point to fill in any gaps in your knowledge. If you have some experience of tinkering with hardware, building mechanisms, or soldering wires and electronic components, you are perfectly equipped for this book. The book assumes that you have used a solderless breadboard before to connect electronic components.

| Software/hardware covered in the book | Operating system requirements |
|---|---|
| Arduino IDE 2.0.3 (or higher) | Windows, macOS, or Linux |
| Arduino Uno or Mega2560 | |

*All you need to start working with this book in addition to the free Arduino IDE software is an Arduino Uno or Arduino Mega2560 board, a solderless breadboard, and a few jumper wires. We will explore many additional, inexpensive hardware components in the individual chapters, but you neither need all of them, nor the exact models that are used in the book (with a few exceptions). If you want to recreate the experiments in a certain chapter, the chapter will point you to the hardware you need.*

**If you are using the digital version of this book, we advise you to type the code yourself or access the code from the book's GitHub repository (a link is available in the next section). Doing so will help you avoid any potential errors related to the copying and pasting of code.**

*This book will enable you to use a wide variety of hardware components for Arduino DIY robots, far beyond the exact parts used in the examples throughout the chapters.*

# Download the example code files

You can download the example code files for this book from GitHub at `https://github.com/PacktPublishing/Practical-Arduino-Robotics`. If there's an update to the code, it will be updated in the GitHub repository.

We also have other code bundles from our rich catalog of books and videos available at `https://github.com/PacktPublishing/`. Check them out!

# Download the color images

We also provide a PDF file that has color images of the screenshots and diagrams used in this book. You can download it here: `https://packt.link/THXao`.

# Conventions used

There are a number of text conventions used throughout this book.

`Code in text`: Indicates code words in text, database table names, folder names, filenames, file extensions, pathnames, dummy URLs, user input, and Twitter handles. Here is an example: "To control the three colors of LEDs with variable brightness via software, we can use the `Dimmer` class that we developed earlier and simply instantiate one instance per color: `redDimmer`, `greenDimmer`, and `blueDimmer` on three different pins."

A block of code is set as follows:

```
enum Mode {
  MANUAL_CONTROL,
  FOLLOW_LINE
};
Mode mode = MANUAL_CONTROL;
```

When we wish to draw your attention to a particular part of a code block, the relevant lines or items are set in bold:

```
if (millis() - last_blink_time >= blink_interval) {
  // Update the last blink time.
  last_blink_time += blink_interval;
  // Execute the blink task.
  blink_task();
}
```

**Bold**: Indicates a new term, an important word, or words that you see onscreen. For instance, words in menus or dialog boxes appear in **bold**. Here is an example: "If you are looking for a specific library, open the library manager by clicking **Tools** | **Manage Libraries...** and use the search bar to find the library you are looking for. If it is available, simply hover your mouse over it and click the **Install** button."

> **Tips or important notes**
> Appear like this.

## Get in touch

Feedback from our readers is always welcome.

**General feedback**: If you have questions about any aspect of this book, email us at `customercare@packtpub.com` and mention the book title in the subject of your message.

**Errata**: Although we have taken every care to ensure the accuracy of our content, mistakes do happen. If you have found a mistake in this book, we would be grateful if you would report this to us. Please visit `www.packtpub.com/support/errata` and fill in the form.

**Piracy**: If you come across any illegal copies of our works in any form on the internet, we would be grateful if you would provide us with the location address or website name. Please contact us at `copyright@packt.com` with a link to the material.

**If you are interested in becoming an author**: If there is a topic that you have expertise in and you are interested in either writing or contributing to a book, please visit `authors.packtpub.com`.

## Share Your Thoughts

Once you've read *Practical Arduino Robotics*, we'd love to hear your thoughts! Scan the QR code below to go straight to the Amazon review page for this book and share your feedback.

`https://packt.link/r/1804613177`

Your review is important to us and the tech community and will help us make sure we're delivering excellent quality content.

## Download a free PDF copy of this book

Thanks for purchasing this book!

Do you like to read on the go but are unable to carry your print books everywhere? Is your eBook purchase not compatible with the device of your choice?

Don't worry, now with every Packt book you get a DRM-free PDF version of that book at no cost.

Read anywhere, any place, on any device. Search, copy, and paste code from your favorite technical books directly into your application.

The perks don't stop there, you can get exclusive access to discounts, newsletters, and great free content in your inbox daily

Follow these simple steps to get the benefits:

1. Scan the QR code or visit the link below

`https://packt.link/free-ebook/9781804613177`

2. Submit your proof of purchase
3. That's it! We'll send your free PDF and other benefits to your email directly

# Part 1: Selecting the Right Components for Your Robots

This part gives an introduction to robotics and provides a detailed overview of the Arduino Uno board and the Arduino IDE, its setup, and key features. In addition, you will learn about the various types of sensors that you can use to let your Arduino robots perceive the world, their interfaces, and how to use them. This part also covers many types of electric actuators, from RC servos to powerful BLDC motors, that you can use to make your robots move. And finally, it gives you the knowledge you need to select the right Arduino board for your projects.

This part has the following chapters:

- *Chapter 1, Introducing Robotics and the Arduino Ecosystem*
- *Chapter 2, Making Robots Perceive the World with Sensors*
- *Chapter 3, Making Your Robot Move and Interact with the World with Actuators*
- *Chapter 4, Selecting the Right Arduino Board for Your Project*

# 1

# Introducing Robotics and the Arduino Ecosystem

Creating robots means building machines that autonomously interact with the physical world. Robotics lies right at the intersection of three exciting engineering fields: computer science, electrical engineering, and mechanical engineering. Building your own robots can seem daunting, even if you are an expert in any of these three fields already. There is a lot to know before you can really get started. But if you successfully master the fundamental skills that this book teaches you, you will soon be ready to create advanced robots yourself. And from there, the possible applications are endless! You can build robots just for fun and the joy of learning more and more with every project. Or you can use these skills to quickly create a physical prototype of a product idea, build smart tools that help you at home or on the job, and automate, entertain, teach, and educate. You might be able to create amazing student projects or even boost your career with these new skills.

Your robot needs a brain – a computer that can run your programs, process sensor signals, and control motors. There are several options for DIY robots. You can simply use your laptop, but that is a costly and pretty clunky option, and it makes interfacing with low-level hardware and implementing real-time control systems difficult. You can also use the popular Raspberry Pi single-board computer, which is a powerful and affordable platform for many DIY projects. However, for many DIY robot projects, Raspberry Pi is overkill and adds unnecessary layers of complexity. In contrast, using **microcontrollers** as the compute platform is extremely affordable, makes interfacing with almost any hardware easy, and is a great way to learn about low-level programming and real-time systems. For many DIY robotics projects, microcontrollers are the ideal platform, and this is where Arduino comes in!

Arduino is an ecosystem of microcontroller boards, tools, and software building blocks that makes creating your own advanced robots exceptionally easy. The two core elements of this ecosystem are **Arduino boards** and the **Arduino Integrated Development Environment** (IDE). Arduino boards are affordable, capable microcontroller boards that are incredibly easy to use, even if you have never worked with electronics before. These features, combined with their wide availability (thanks to their open source design) have truly made microcontrollers accessible to anyone. The Arduino board will be the brain of your robot, running the program you wrote for it. It will sense the world with sensors and make the motors of your robot move accordingly. The Arduino IDE is what we use to develop the programs for your robots and to transfer them from your computer to the Arduino. The Arduino IDE sits in the sweet spot of microcontroller IDEs between graphical programming languages (drag and drop programming) and highly hardware-specific IDEs provided by microcontroller manufacturers. While the former is easy to master but can be very limiting, the latter gives access to a microcontroller's full potential at the cost of taking years to master. In contrast, the Arduino IDE embodies the 80/20 principle: it gives us access to roughly 80% of the microcontrollers' capabilities with only 20% of the effort.

This chapter will kick-start your journey toward building and programming your own robots by answering the following questions:

- What are the main components of a robot?
- What is an Arduino microcontroller board?
- What is the Arduino ecosystem?
- How do you program your Arduino-based robot?

By the end of this chapter, you will know the answers to all of these questions; you will have successfully set up the Arduino IDE on your computer, and your first Arduino program will be running on your Arduino.

# Technical requirements

You will need an Arduino (or compatible) microcontroller board and the matching USB cable to connect it to your computer. In this chapter, we will assume that you have an Arduino Uno Rev3, a great Arduino board to start with.

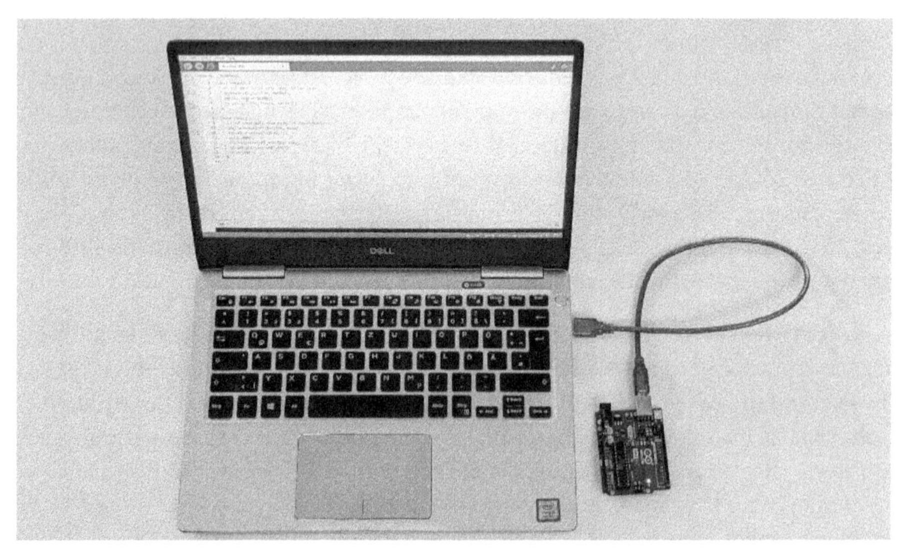

Figure 1.1 – Your computer, a USB cable, and an Arduino Uno are all you need for this chapter

# The main components of a robot

For the purpose of this book, we will define a robot as a machine that can autonomously and intentionally interact with its environment. This definition directly provides the three key capabilities that a robot must have:

- Sense its environment
- Reason about the state of its environment and compute its appropriate reaction
- Physically move to interact with the environment

*Sense-Reason-Act* is a commonly cited robotics paradigm and is visualized in *Figure 1.2*:

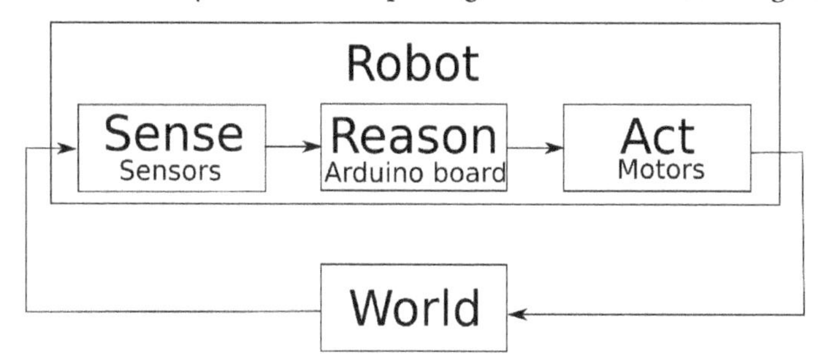


Figure 1.2 – The Sense-Reason-Act paradigm of robotics for Arduino robots

A robot needs to **sense** the aspects of its environment that are relevant to its task. If you build a robot to water your plants, the robot needs to sense the moisture of the soil, for example. If you build an automated transporting robot that needs to follow a line on the floor, it needs to sense the position of the line between its wheels. Choosing the right sensors for your robot is an important step in the design process. This book will introduce you to a variety of readily available and easy-to-use sensor types that work with Arduino robots, along with examples of how to use and integrate them into your robot.

**Reasoning**, based on sensor signals, is the task of the program that you implement on your Arduino. Your Arduino can run programs just like a real computer, although, typically, with much tighter restrictions on computational power and available memory. We will keep these limitations in mind when we write our Arduino programs. In practice, there is often a trade-off between choosing sensors and implementing the control logic. Adding more and better sensors can make controlling your robot easier, but also leads to more complex hardware and higher cost. More advanced software that runs on your robot can often make up for the lack of high-quality sensor data but is more difficult and time-consuming to develop. An example of this is the combination of the human eye and brain: the eye is not a very good image sensor. It has a blind spot, variable resolution and color sensitivity across its field of view, and a less-than-ideal lens. Yet our brain turns this all but perfect sensor data into crystal clear images under a wide variety of conditions (enabling you to read this book, for example). Here, the capabilities of the *software* in our brain make up for the rather poor sensor. In contrast, most robots that heavily rely on image data for their tasks (autonomous cars, for example) have very good camera sensors, and they can use this image data without much post-processing.

Your robot will need motors (or more generally, *actuators*) to physically **act**. These can be electric motors, pneumatic or hydraulic cylinders, gasoline engines, or even rocket thrusters. In this book, we will focus our attention on electric motors, but let's keep in mind that there are many more actuator types to explore. In contrast to sensors that can often be directly connected to an Arduino and even powered from it, motors typically need some form of driver (or amplifier) in between the Arduino and the motor. This is because motors tend to require much more power than an Arduino alone can provide. You can imagine the motor as a big and complex piece of machinery, and the driver electronics as the specialist operator of this piece of machinery. The Arduino will communicate with the driver/operator to tell them what the machine should be doing, but it leaves the actual operation of the machine to the operator. There is a similar trade-off between the quality and capabilities of your robot's actuators and the complexity of the robot program that we mentioned earlier with the sensors. It is generally easier to work with very precise, fast, and strong actuators, but more sophisticated software can often make up for imperfect actuators. The human brain is again a good example: humans can perform incredibly precise and dynamic movements, even though there is quite a bit of latency in our nervous system. The brain's *software* can make up for the shortcomings of the hardware.

## What different types of robots are there?

There are many types of robots that fall under our broad definition of a robot. The content of this book is relevant for all of them, so no matter whether you want to build a flying robot or your own 3D printer, this book will be very useful for you. In fact, there are so many different kinds and types of robots built for a diverse range of applications that it is not at all straightforward to classify them consistently. At a very high level, it is common to differentiate between **mobile robots** that can move around in the world and **stationary robots**. Prominent examples of mobile robots are self-driving cars, and prominent examples of stationary robots are robotic arms widely used in industrial manufacturing, for example, to weld car parts together.

Mobile robots can be divided into subcategories based on whether they are ground-based, flying, or swimming, how they propel themselves (for example, with wheels or legs), and whether they have the means to manipulate their environment with an arm and a gripper. Using these categories, you could describe a humanoid robot (one that looks like a human) as a ground-based, legged, dual-arm, mobile manipulator.

In this book, we will primarily use ground-based, wheeled, mobile robots as examples. This type of robot is easy and inexpensive to build, can be tested virtually everywhere, and exemplifies the key concepts that apply to almost every category of robots.

# What is an Arduino microcontroller board?

The Arduino ecosystem is a set of microcontroller hardware, software tools, and libraries that make programming microcontrollers much more accessible than it has been traditionally. In this section, we will discuss what microcontrollers are, and how Arduino helps us use them easily and effectively.

## What is a microcontroller?

A microcontroller is an **integrated circuit** (**IC**) that contains all the necessary components to make it a little standalone computer. An IC is a complex piece of electrical circuitry that is made from a single, tiny piece of silicon (a **chip**). A microcontroller has a processor (CPU), memory and storage, and usually a host of other hardware peripherals that implement standardized low-level interfaces to other electrical circuits. These interfaces can be used to communicate with other pieces of hardware, such as sensors and motor drivers. To make them easier to work with, manufacturers put the microcontroller IC in a housing that is much larger than the IC itself. The housing exposes metal pins that are connected to the electrical contacts of the IC inside. The terms *IC* and *chip* are often used interchangeably to describe the combination of the silicon chip and its housing.

The simplest interface is digital **input/output** (**I/O**). Most of the pins of a microcontroller can be configured to be either digital inputs or output, and our program can either set their voltage level low (0V) or high (typically 3.3V or 5V) or read their voltage level (as digital LOW or HIGH). We will use this functionality, for example, to control LEDs that are connected to a digital output and to read the status of switches that are connected to a digital input. While digital I/O is extremely useful to transmit the state of a single bit, it is not a good interface for the transmission of a lot of data. For this task, slightly more complicated interfaces (often called **buses**) are used. A well-known example of a hardware interface for data transmission is the **Universal Serial Bus** (**USB**). This is a very powerful, flexible, but complicated-to-implement interface. It is usually not directly used in DIY robotics. The interfaces we use for our robots are simpler and easier to understand and use. Interfaces commonly used in Arduino robots include **Universal Asynchronous Receiver/Transmitter** (**UART**), **Serial Peripheral Interface** (**SPI**), the **Inter-Integrated Circuit** (**I2C**) bus, **Pulse Width Modulation** (**PWM**), and analog data transmission.

A microcontroller will have dedicated hardware for some or even all of these interfaces integrated on the same chip. That means the CPU does not need to spend time handling the transmission or reception of data (and we do not need to spend time implementing any of the algorithms for these interfaces). Our program will simply tell the CPU to pass data to these peripherals for transmission, or ask the peripheral for any received data. Easy access to these hardware interfaces is a major reason why microcontrollers are so great for DIY robotics. Most sensors, motor drivers, or any other component we might want to use in our robot will have one of these interfaces, and as long as we know how to program a microcontroller, we can easily *talk* to these components. More generally, access to these interfaces makes microcontrollers great for any kind of *hardware hacking*. Most chips that you find in appliances, toys, tools, printers, and game console controllers (everywhere, really) likely communicate over one of these interfaces. You can use a microcontroller to *talk* to them directly and gain lower-level access to their capabilities, rather than having to use the interfaces that the manufacturer exposes. An example of this (that works with very little hacking) is the popular use of the Nintendo Wii Nunchuck controller, a nice and inexpensive in-hand joystick, in DIY robotics projects. It uses the I2C interface.

### Programming microcontrollers

Writing programs for microcontrollers is different from more typical programming on PCs. One of the main differences is that, in general, there is no **operating system** (**OS**) on a microcontroller. Your program is the only thing running on the CPU, and there is no OS to help manage resources or prevent the system from crashing. Your program also has direct access to hardware functionalities, such as sending power to any of the digital output pins. This is good in many ways: it makes it easy to understand *what is going on* during program execution, it means that there is almost no computational overhead from anything that is not your program, and it makes hardware access very transparent. However, it also means that we need to be a little more aware of what can go wrong if our program does not behave as intended since it is easy to make the program stall (for example, if we wait for a condition that never occurs), crash, or even damage attached hardware if we sent incorrect output signals. This direct access to hardware functionalities without the safeguards and layers of abstraction that an OS provides is why we sometimes call this type of programming *hardware-near* programming or, more commonly, *embedded* programming. Some languages lend themselves better to embedded programming than others. Most commonly, microcontrollers are programmed using the C language since it offers powerful and elegant ways to write highly efficient code that lets the programmers squeeze out every bit of performance, even from the smallest microcontroller. Even though C is sometimes regarded as a low-level language, is still much more user-friendly than going one level closer to machine language and programming in *assembly*.

The Arduino IDE supports not only C but also C++ (often abbreviated together as C/C++), and we rely on C++ to enable **object-oriented programming** (**OOP**), which allows us to easily use many third-party software libraries and make our own code reusable across different projects. We will take a much closer look at this in a later chapter. All examples in this book are written in C/C++.

The Arduino IDE does not support any other programming language, such as Python, for example. However, the Arduino project officially supports MicroPython, and you can use MicroPython and the OpenMV IDE to program Arduino boards using the popular Python programming language.

### Pain points of working with microcontrollers

The lack of hardware abstraction provided by an OS means that the exact commands for doing a certain task (for example, setting a certain output pin high or sending data over a certain interface) can be very different from microcontroller to microcontroller, and you, the programmer, often have to dig through hundreds of pages of datasheets to understand how exactly to perform a certain operation, such as sending data over I2C on a new microcontroller. Even translating your final program to the machine code that can run on the microcontroller (**compiling**) can be complicated and extremely specific to the exact microcontroller you are using. And finally, transmitting the compiled program to the microcontroller often requires specialized hardware that can be complicated to use. For these reasons, the barriers to working with microcontrollers have traditionally been high, and they were mostly used by expert engineers rather than hobbyists, artists, students, or makers. To summarize, the traditional pain points preventing more people from using microcontrollers are as follows:

- The fact that programs are highly hardware specific
- The need for specialized compilers
- The need for specialized hardware to transmit the compiled program

The creators of Arduino had the goal to get rid of these barriers and enable non-experts to use microcontrollers in their projects. Arduino is not the only project with this goal, but it has been the most successful, and it has had the largest impact on the maker scene by enabling many more people to take advantage of the power of microcontrollers.

## What exactly is Arduino?

Arduino is a company that makes and distributes open source hardware and software. However, when we talk about *an Arduino* that controls a robot, we usually mean a circuit board with an Arduino-compatible microcontroller on it. But, as we already mentioned, Arduino is more than the physical hardware: it is an entire ecosystem of software, tools, and hardware that makes using microcontrollers accessible for anyone, with Arduino boards and the Arduino IDE at its center. Together, the IDE and the supported boards address the pain points we identified earlier. They simplify and streamline writing programs, compile them, and transmit the compiled programs to the board. There are also extension boards (called Arduino *shields*) that add additional functionality to standard Arduino boards without requiring specialized electronics skills.

There is a variety of different supported Arduino boards, and we will look at their differences and distinctions in more detail in a later chapter. One great aspect of Arduino is that the workflow of writing programs and transmitting them to the physical board is the same across all the different boards. In fact, in many cases, the exact same code can be used across different boards with different microcontrollers. No need to comb through datasheets for specific register names or to use specialized hardware for programming. When everything is set up, all you need to do is to connect your Arduino board to your computer with a standard USB cable, click one button, and the program will be compiled, transmitted, and starts running on your board automatically. For the remainder of this chapter, we will assume that our Arduino board is an Arduino Uno Rev3, but the steps to program it would *not be different* from any other official Arduino board that is supported by the IDE. Arduino boards also contain all the components needed to support the microcontroller and provide easy access to all of its pins. So, to answer the question in the title of this section: an Arduino microcontroller board is an easy-to-use circuit board with a microcontroller and all the electronics to support it, and can be programmed using the Arduino IDE.

Because of the popularity of Arduino, the fact that both hardware and software are open source, and the use of licenses that allow reusing the hardware design in commercial projects, a huge number of third-party projects that expand the Arduino ecosystem have emerged. There are numerous manufacturers of Arduino shields or entire (robot) starter kits, and even **original equipment manufacturers** (**OEMs**) of electronics components make development boards intended to be integrated into Arduino-based projects. There is also an incredible number of online tutorials, YouTube videos, and *How-To* guides that show you how to build a certain project with Arduino. As exciting as all of this is, experience shows that it can be overwhelming for anyone who is just entering the field of DIY electronics and robotics. And without a solid understanding of the basics, much of this content can feel more like noise rather than educational. The goal of this book is to equip you with a solid understanding of robotics with Arduino, the possibilities and limitations, as well as the pros and cons of many different design choices you have to make when building your own robot. After you have read this book, you will know where to start any Arduino robotics project that you want to build, and you will be able to understand and make use of the large amount of information and resources that are available from and for the Arduino robotics community.

# The Arduino ecosystem

Let's now take a look at the major components of both the Arduino IDE and the Arduino Uno board. After that, we will be ready to set everything up and write, transmit, and run our first Arduino program.

## The Arduino IDE

The Arduino IDE is free and open source software available for Windows, macOS, and Linux systems. Arduino also offers a web-based version that works in your web browser and stores your files in the cloud as opposed to locally. The examples in this book assume that you are using a locally installed version of the IDE on a Windows computer, but most steps will be identical, or at least very similar, across the three systems. In September 2022, the Arduino IDE received a major upgrade from version 1.8 to version 2.0. Version 2.0 is easier to use and more powerful, which means that now is a better time than ever to start using Arduino!. *Figure 1.3* shows how the Arduino IDE 2.0.3 looks on startup:

Figure 1.3 – The Arduino IDE right after launch

We will take a closer look at the Arduino IDE later in this chapter after we have set it up, but for now, let us understand what the key components and the purpose of this IDE are:

- A **text editor** lets you write your C/C++ program right in the IDE. In Arduino parlance, programs are called **sketches**, a nod to the IDE's origins in the *Processing* project (`https://processing.org/`). The editor has some useful features, such as syntax highlighting, auto-formatting, and auto-completion. However, it is a pretty simple editor still. While this simplicity is intentional, it is not as powerful as other modern code editors, such as Visual Studio Code, for example. There are more advanced alternatives to the built-in text editor, but for the purpose of this book, we will be using the IDE's text editor as it is the most seamless option to develop Arduino programs.
- A **compiler** lets you translate your code to machine language for the microcontroller you are using. Arduino makes using the compiler very easy; all it requires is one click on the checkmark button in the top-left corner. When the compiler runs, it produces some output in the console window that appears under the editor window. This output can be very useful for troubleshooting your program.

- A **programmer** lets you seamlessly transfer the compiled machine code to your Arduino board that is connected to your PC via USB. All it takes is, again, one click on the arrow symbol in the top-left corner. And just like the compiler, the programmer will generate some output in the console under the editor that is useful for troubleshooting if the programming does not work as expected.
- A **debugger** lets you run your code step by step on the Arduino's microcontroller. This advanced feature can be very useful for finding problems that happen during program execution. However, the debugger only works with certain boards (not with the Arduino Uno) and may require additional hardware, depending on the board. If a compatible board is connected, the Play button with the little bug in the top-left corner of the Arduino IDE becomes available to start a debugging session.
- A **serial monitor** lets you send text from the PC to the Arduino and can display text that your Arduino sends to the PC. This works with all boards and is an incredibly useful tool for debugging and interacting with your running program. The IDE also provides a **serial plotter** that can display multiple lines of numeric data sent from your Arduino board as a 2D live plot. The buttons to start the serial plotter and the serial monitor are located in the top-right corner of the IDE.
- Built-in **hardware abstraction** libraries make your code more or less independent of the board you are working with. This feature is not really visible on the surface, but it might be the most important one. This is what allows you to program even complex Arduino boards without having to look at the datasheet of the microcontroller, and to use the exact same code that you developed for one board across many different Arduino boards.

Now that we have a good understanding of what the Arduino IDE is, let's look at the other aspect of the Arduino ecosystem: one of the boards that can be programmed with the Arduino IDE.

## The Arduino Uno

The **Arduino Uno** is the most widely used Arduino board due to its simplicity, ruggedness, and widespread availability. It is built around the ATmega328P microcontroller, which has a long history in low-cost DIY electronics projects. The Uno is a great board to get started with Arduino as it is low-cost and simple, yet it contains all the key functionalities of a microcontroller board and is very forgiving to wiring and other mistakes that might damage more complicated boards. Similar to what we did for the IDE, let us look at the key components of the board and what their functionalities are. We will go a little more into detail here since we will rely on a solid understanding of the board components in future chapters. *Figure 1.4* shows the Arduino Uno R3 board for reference:

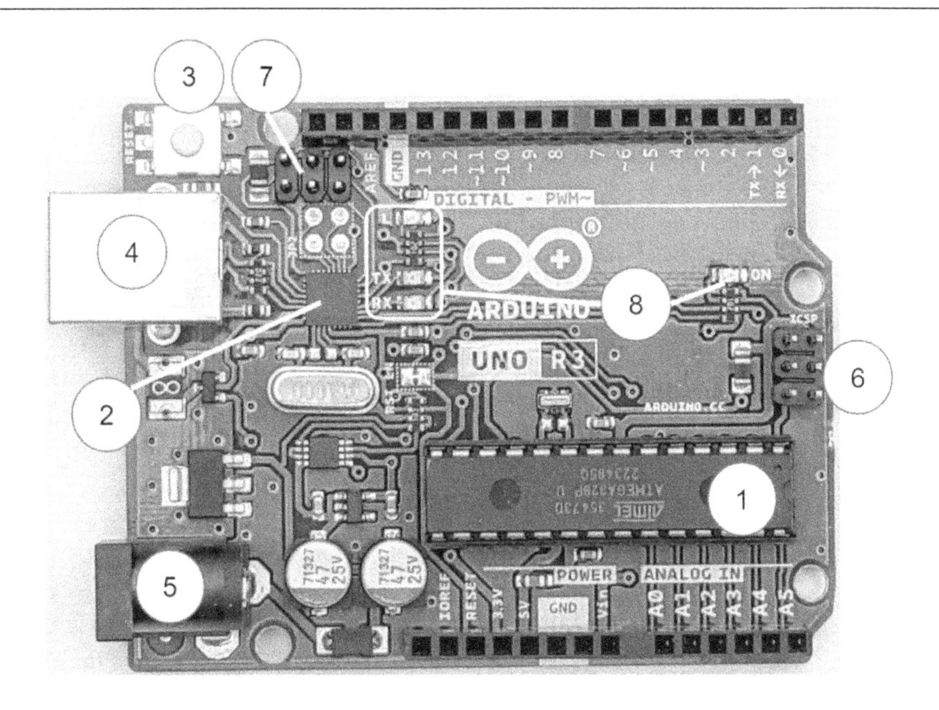


Figure 1.4 – The Arduino Uno Rev3 microcontroller board

The key components that we will discuss next are labeled with numbers.

### *Main electronic components*

The biggest chip on the board is the **ATmega328P** microcontroller (**1**). It is so big because it is enclosed in a socketed **Dual Inline Package** (**DIP**) housing, which allows you to replace it in case it gets damaged. This is the microcontroller that we will be programming toward the end of this chapter, and that controls all the output pins that you see lined along the bottom and top edges of the board. Let us call it the **main microcontroller**. It only needs very few peripheral components to run, namely the two small ceramic capacitors under and above it, and the little oscillator above it (under the **R3** marking).

The second-largest chip on the board is another microcontroller, an ATmega16U2 microcontroller in a much smaller package that is soldered directly to the board (**2**). It comes pre-programmed with special firmware that lets it act as an interface between the USB connector and the ATmega328P. More technically, it implements a USB-to-serial converter. Let's call this one the **programmer**. When we click the program button in the IDE, our computer starts communicating with this chip over USB, transmitting the compiled machine code to it. The programmer receives the code byte-by-byte over USB and transmits it to the main microcontroller over a serial interface. To enable this type of programming, there is a special bit of software already programmed into the main microcontroller, called the **bootloader**. We do not need to go into any more detail here about how the bootloader works. But it is worth mentioning that if you ever want to replace the main microcontroller, you need to either purchase one that is already pre-programmed with the Arduino bootloader or flash the Arduino bootloader onto it yourself.

There is a small **RESET** button in the top-left corner of the board (**3**). Pressing this button causes the main microcontroller to reset. It will stop the current program execution and start executing the program from the start, similar to rebooting a PC. The reboot of a microcontroller happens almost instantly. The reset button is typically not used very often, but it can be useful in the development process when you want to run your program from the start again to test something, or when the program has unexpectedly halted, and you need to reset it.

Below the reset button is the **USB connector** (**4**). On the Uno, this is a male USB Type-B port. Although this is a dated connector by now, it has the advantage of being very rugged and hard to damage. Smaller or third-party Arduino boards often have more modern Mini, Micro, or Type-C USB connectors.

> **Note**
>
> The Arduino can run on power provided by the USB port. However, if you use the Arduino's 5V to power other consumers such as motors or long LED strips, you might exceed the 2.5 W of power that a USB port can provide. In this case, your PC will likely shut down the power to this port to protect itself from overcurrent. In the worst case though, damage to the PC's USB port can occur.

The **power connector** for external power as well as the board's power system is in the bottom-left corner (**5**). You can power the board through this connector with a voltage ranging from 7V to 12V, from a 2-cell lithium battery, for example. Directly above this connector is the *linear* voltage regulator that turns this input voltage into 5V, which is needed to run the microcontrollers. It is worth noting that this is not a very efficient kind of converter and can get hot quickly, depending on the power requirement of your project. In general, it is better practice to use a more efficient *switched* 5V regulator between the battery and your Arduino and connect its 5V output directly to the pin on your Arduino that is marked **5V**. This way, you circumvent the onboard linear regulator altogether. We will investigate this and other power considerations in more detail in a later chapter.

There are **2 male 3x2 pin headers** on the board. The one on the right (**6**) is labeled **ICSP**, which stands for **In-Circuit Serial Programming**. This is primarily a low-level programming port that can be used to flash the bootloader onto the main microcontroller. Beyond that, it can be used to access the SPI interface of the main microcontroller. The other male pin header is the ICSP interface for the programmer microcontroller (**7**). This is used to flash the USB-to-serial firmware onto the programmer during the production of the board. We will not need to use this interface.

There are **two rows of female pin headers** along the top and the bottom edge of the board. These pins are your Arduino's interface to the world! This is where we connect all the physical components such as sensors, motor drivers, LEDs, displays, buttons, or potentiometers that we want to interact with. Very conveniently, the pins are all grouped by their function and labeled, making it easy for us to reference specific pins in our code (without having to look up anything in the datasheet of the microcontroller). Let us now look at the various groups of pins and their functions.

### *Power pins*

As the name suggests, the **POWER** pins have to do with the power system of the board rather than with logical inputs and outputs. Going from left to right, they are as follows:

- **IOREF**: This is a 5V pin that is meant to signal to specialized connected boards (called **Arduino shields**) what voltage the Arduino board is running on.
- **RESET**: This pin is connected to the reset line of the microcontroller, which is internally pulled high by a large resistor. Connecting this pin to GND has the same effect as pressing the RESET button, effectively rebooting the main microcontroller.
- **3.3V**: This is an output pin that provides 3.3V from a very small internal linear regulator. It can be used to power small peripherals that require 3.3V, but it is again better practice to not rely on it but to use an external switched 3.3V regulator instead.
- **5V**: This pin connects directly to the 5V net of the board. It can be used to tap off 5V power for other components of your robot when the Uno is powered by the USB connector or the external power input. Or it can be used to power the Uno from a 5V source, such as an external 5V regulator.
- **GND**: These pins provide access to the board reference potential (ground). In general, every component that we want to connect to our Arduino needs to share the same reference potential and must therefore be connected to the Arduino's GND net. This is why the Uno has three GND pins. In practice, though, this is often not enough, and we will learn how we can use a solderless breadboard to get even better access to GND.
- **Vin**: This pin is directly connected to the barrel connector in the bottom-left corner of the board. It can be used for the same purpose – to power the board from a battery or other voltage source in the range from 7V to 12V (with the caveat of relying on the inefficient onboard linear regulator).

On the left side of the top pin header is the **AREF** pin. It can be used to feed in an external reference voltage to Arduino Uno's built-in **analog-to-digital converter** (**ADC**). By default, the ADC uses 5V as the voltage reference, which is acceptable for most applications, so the AREF pin is rarely used in practice.

### *GPIO pins*

All other pins that are neither POWER pins nor AREF are **General Purpose Input/Output** (**GPIO**) pins and can be used as **DIGITAL** inputs and outputs. However, most pins of a microcontroller can be used for more than one functionality, depending on their configuration. To avoid confusion and simplify things a little, only pins 0 to 13 on the Arduino are labeled as **DIGITAL**. These digital pins can be used to either read digital input signals as HIGH or LOW (for example, from a push button), or to send a single bit of information (for example, to control a LED) by setting their output voltage to HIGH (5V) or LOW (GND). Some of the DIGITAL pins have a tilde in front of their pin number. This indicates that they can be used to send a PWM signal. This means we can ask them to automatically go HIGH and LOW repeatedly at a very high frequency, and we can easily control the width of the high pulse (from 0%, always LOW, to 100%, always HIGH). This capability is useful to control the brightness of LEDs, the speed of DC motors, or the position of RC servos, for example. Lastly, pins 0 and 1 are marked as **RX** and **TX**. This is because they are used by the serial interface between the main microcontroller and the programmer. The same interface is also used for any other communications with the IDE (for example, the serial monitor that we will look at later in this chapter). It is generally a good idea to try to avoid using these pins for anything else to avoid interference with these functionalities. If you need an additional serial interface for your application, you can use Arduino's software serial to turn any two digital pins into an additional set of TX and RX pins.

The **ANALOG IN** pins can be used to read analog signals in the range of 0V to 5V. While digital inputs can only distinguish between two states (HIGH and LOW), the analog input pins can read voltages with a resolution of 10 bit, or 1,024 steps. This means they read a value of `0` when connected to GND, `512` when connected to 2.5V, and `1023` when connected to 5V, for example. A common use case for analog input pins is to read the output voltage of a potentiometer that can be used as a dial to set a continuous input value. These pins can alternatively be used as digital pins 14 through 19. Pins A4 and A5 are broken out twice on the Arduino board: once on the bottom right of the board and a second time, unlabeled, on the top left.

Beyond these basic functions (power, digital, and analog), many of the pins have alternative functions that can be selected via software. Let us get a quick overview of these additional functionalities here, even though we will take a much closer look (with examples) at how to use them in the later chapters. The Arduino IDE with its built-in hardware abstraction libraries makes configuring these special functionalities extremely easy:

- The two analog pins A4 and A5 (and also the two unlabeled pins in the top-left corner) can be configured as the two pins for the *I2C* bus interface, namely SDA (pin A4) and SCL (pin A5).
- Pins 11, 12, and 13 can be configured to be the three pins of the *SPI* interface, MOSI, MISO, and SCK, respectively. In this case, pin 10 can be used as a hardware-controlled chip select pin for the SPI bus, but any software-controlled digital pin can also serve this function.
- Pins 2 and 3 can be configured as hardware *interrupt* triggers. This is a useful capability to react to changes in the voltage level on these pins (from LOW to HIGH or HIGH to LOW) with the shortest possible latency. Example applications for this include decoding the PWM signal from an RC receiver or counting the steps of an incremental motor encoder.

Our overview of the Arduino Uno is almost complete, but no microcontroller board is complete without a few indicator LEDs, and the Uno is no exception.

### *LEDs*

The last major components to mention are the three **yellow status LEDs** and the **green power LED** on the board (**8**). Let us look at the three yellow status LEDs that are all arranged in a vertical line under pin 13, labeled **L**, **TX**, and **RX**:

- The LED labeled **L** is connected to digital pin 13. It lights up when the pin is HIGH (5V), and it is off when the pin is LOW (GND). It is quite common to at least have one status LED on your robot to signal its current state. In the simplest case, you can let this LED constantly blink a *heartbeat* signal to show that the program is still running. When the LED is not blinking, you can immediately see that something unexpected has happened and that the program is stalled. Having this status LED built right into the board is very useful, and the very first program we will write in this chapter will not do much more than let this LED blink.
- The two LEDs labeled **TX** and **RX** signal activity on the main microcontroller's serial interface that is used to communicate with the programmer microcontroller and, by extension, with your PC. They are not directly connected to the digital pins 0 and 1 but are driven by the separate pins of the programmer. Whenever the main microcontroller is receiving data from the PC, the RX LED lights up, and whenever the main microcontroller sends data to the PC, the TX LED lights up. These LEDs are especially useful during the programming process to verify that the program is in fact transmitted from the PC to the board, in which case, both will blink rapidly for a brief period.

The green LED labeled **ON** is located on the right side of the board. When the board has good 5V power, this LED will be solidly on. It is useful for troubleshooting power issues with your project. If this LED starts flickering, shines less bright than usual, or even switches off altogether, you know that something is wrong with the power that is supplied to the board.

You now have a detailed understanding of the hardware components and functionalities of the Arduino Uno board. We will build on this knowledge in later chapters, where we look at many of these functionalities in more detail and introduce the code that is necessary to use them. In the next section of this chapter, we will set up the Arduino IDE on your PC and write our first program that will let the LED attached to pin 13 blink. We will also transmit a message over the serial port from the Arduino to the PC.

## How to program your Arduino board

The main goal of the Arduino project is to make programming microcontrollers easy, and this section will walk you through the necessary steps. If this is your first time working with microcontrollers, you will hopefully enjoy the process. If you have experience in programming microcontrollers in more traditional ways with hardware-specific IDEs, ISP programmers, and a datasheet always on hand, you might be surprised by how much smoother the experience with the Arduino toolchain is. The whole process involves four steps, and we will go through each of them:

- Installing the Arduino IDE on your system
- Writing the program code in the IDE's editor
- Compiling the code for your Arduino board
- Transmitting the compiled machine code to your Arduino board

## Installing the Arduino IDE

To download the Arduino IDE, go to `https://www.arduino.cc/en/software` and select the download link for your operating system (Windows, Linux, and macOS are supported). Download the installer and follow the instructions for the full installation. The installation includes the hardware drivers that are necessary to enable the IDE to communicate with the Arduino boards via USB, so make sure to select the option to install the hardware drivers when prompted.

Once the IDE is installed successfully, you can open it, and it will look just as shown in *Figure 1.3*. The IDE has a simple and clear interface. Besides the buttons we already discussed, there are several drop-down menus accessible at the top and a few more shortcut buttons on the left. These include access to the Boards Manager and Library Manager, which we will talk about in more detail later.

You will notice that the IDE already shows some code in the editor when you first open it. This is a bare minimum, valid Arduino program (or sketch) that can be compiled and run on any Arduino board, but that does nothing at all.

You can see that every Arduino program or sketch consists of at least two functions: `setup()` and `loop()`. Just like the comments suggest, the `setup` function is executed only once right after the program starts. In it, we typically put all the code that configures the functionality of the pins and that initializes connected hardware, such as the sensors of our robot. After the `setup` function has been completed, the program will repeatedly execute the `loop` function as fast as it possibly can, until the Arduino is shut off or reset. Let us fill in these empty functions with a bit of code that actually does something! Even with only the Arduino Uno board and no additional hardware to interface with, we can still use the built-in LED on pin 13 and the serial interface to generate some interesting behavior and bring our Arduino board to life.

## Writing your first Arduino program

To use the LED that is connected to pin 13, we must first configure pin 13 as a digital output pin. This will enable it to drive the LED. To do this, we need to add the following line of code to the `setup` function:

```
pinMode(13, OUTPUT);
```

This will tell the microcontroller on our Arduino board that we want to use this pin as an output, and not as input or as part of the SPI interface, for example. This line of code will work for our Arduino Uno where the LED is connected to pin 13. But it would not produce the intended effect on a different board type where the LED is connected to a different pin. We can instead write the following line, where we replaced the number 13 with the `LED_BUILTIN` macro:

```
pinMode(LED_BUILTIN, OUTPUT);
```

During compilation, `LED_BUILTIN` will be automatically replaced with the correct pin number for the board we are working with, which means that our program will work across many different boards. Similarly, `OUTPUT` will be replaced with the number `1`.

To enable the serial interface such that we can send messages between the PC and the Arduino, we need to add the following line of code to the `setup` function:

```
Serial.begin(115200);
```

This line will start the serial interface at a baud rate of `115200` bps. The baud rate determines the data transmission speed of the serial interface. 115,200 bps will be our default serial baud rate for all examples in this book, as it allows for both reliable and fast communication. Directly after this line, we can start to use the serial interface. As is customary in programming tutorials, let us let our program print the words `"Hello, world!"` to the serial monitor. To do this, all we need to do is add the following line at the end of our `setup` function:

```
Serial.println("Hello, world!");
```

The `Serial.println()` functions will write the characters of the sentence `"Hello, world!"` one by one to the serial output. It will automatically add a line break at the end of it to make parsing and displaying this message easier. This is it for the `setup` function; there is nothing else we need to add.

In the `loop` function, we will periodically turn the LED on and off again, and also print out a corresponding message. To turn the LED on, we need to add the following line at the top of the `loop` function:

```
digitalWrite(LED_BUILTIN, HIGH);
```

This line sets the digital output of pin 13 to 5V, causing the LED to light up. Let's also print a corresponding message and wait for half a second before we turn the LED off again with the following two lines:

```
Serial.println("LED on. ");
delay(500);
```

The `delay` function implements a *busy wait* and will halt the program execution for the specified number of milliseconds. It is usually good practice to avoid the use of `delay()` (or any other form of busy waiting in general) everywhere except in the `setup` function. The reason is that a busy wait prevents the CPU from doing anything useful during this time, which can cause many unintended consequences. We will introduce methods to avoid the use of delay in a later chapter, but for this first example, there is nothing wrong with using it. Lastly, let us turn the LED off again, print a notification, and wait another half a second before we automatically repeat the blink process all over again:

```
digitalWrite(LED_BUILTIN, LOW);
Serial.println("LED off.");
delay(500);
```

And this is all we need to do to write a program that lets the LED blink at a frequency of 1 Hz and outputs a little notification over the serial port every time it turns the LED on or off. *Figure 1.5* shows how the final program looks in the Arduino IDE:

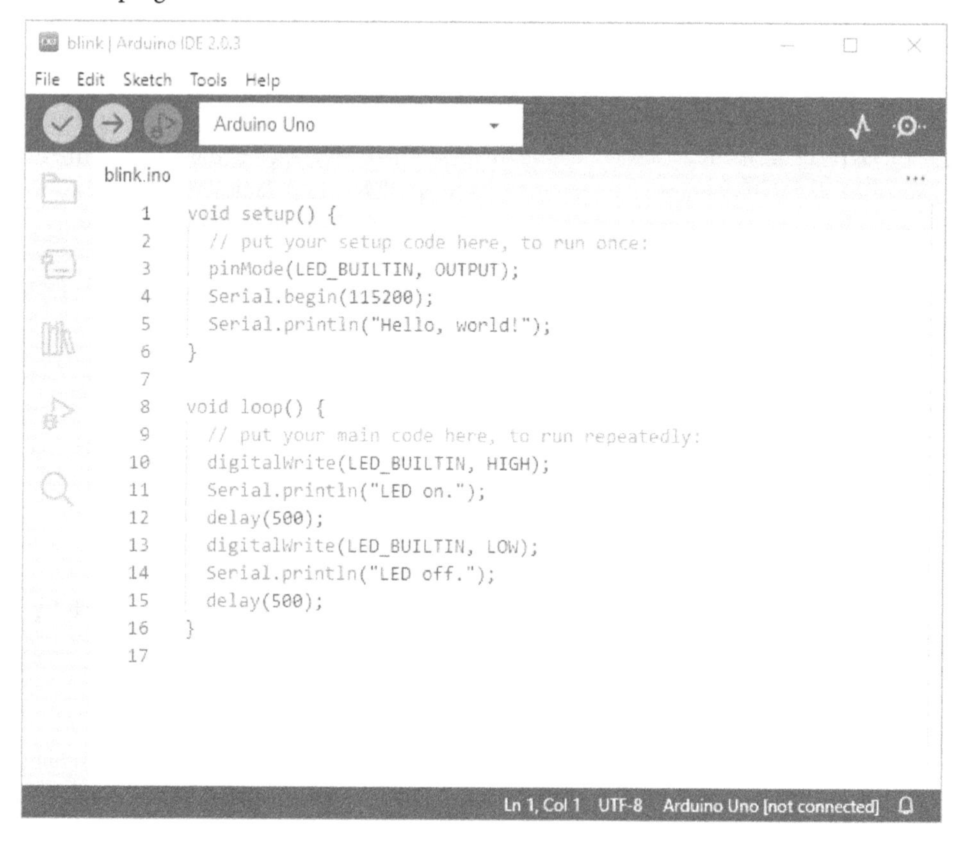


Figure 1.5 – The blink program with serial output statements

You might have noticed that this is not actually valid C++ code, since the `main()` function is missing. During the compilation process, this function will be automatically generated, and it calls the `setup` function once and then the `loop` function in an endless loop.

## Compiling the code for your Arduino board

Now that you have written the code for your program, we need to translate it into machine code for the microcontroller on the Arduino Uno. To do this, you first need to tell the IDE what board you are using. Click the drop-down window in the center, click on **Select other board and port** and type the word `uno` in the search bar under **BOARDS**. Select the **Arduino Uno** entry. If your Arduino is connected to your PC, you can find and select the serial port on the right under **PORTS**. This step is needed to transmit the program to the board. *Figure 1.6* shows where to select the board type and the serial port that you will be using for programming. When you are done, click **OK**:

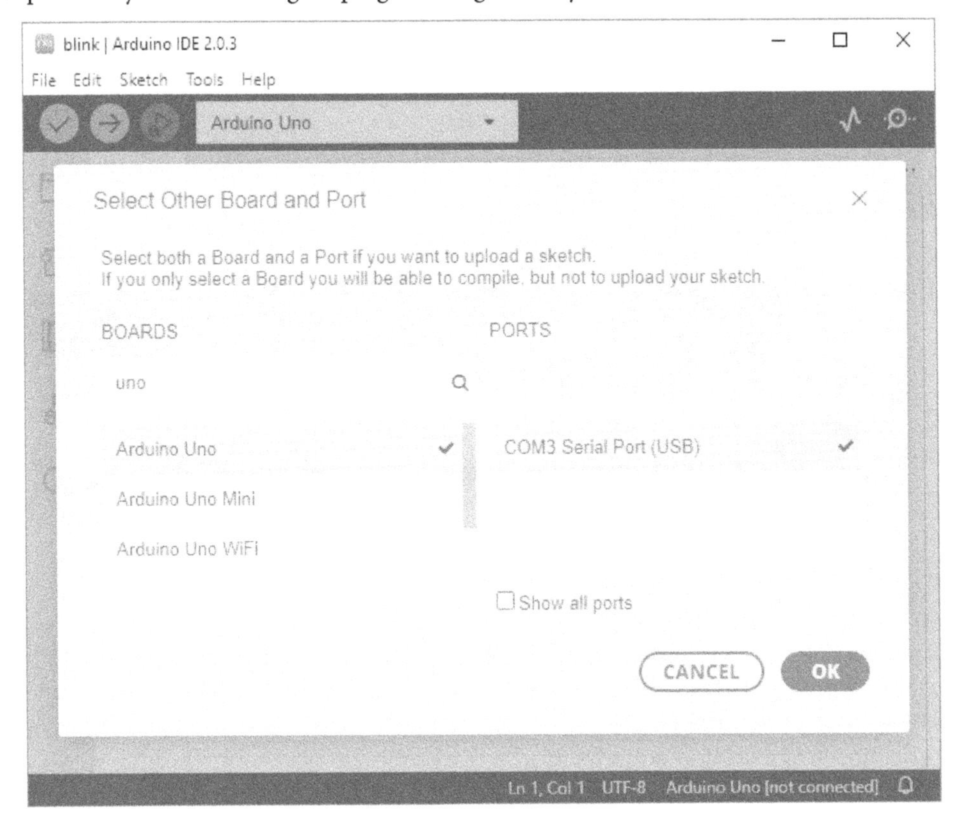


Figure 1.6 – Selecting the board type and serial port

Now, you can click on the checkmark symbol in the top-left corner to compile your program. As soon as you click on it, the console window (titled **Output**) will appear under the editor. The IDE will also display a pop-up message that says **Compiling sketch...**, and if all goes well, this message will change to **Done compiling**. A message in the black console window under the editor will inform you how much program storage space and dynamic memory are being used by your program. *Figure 1.7* shows what the IDE looks like when compilation is successful:

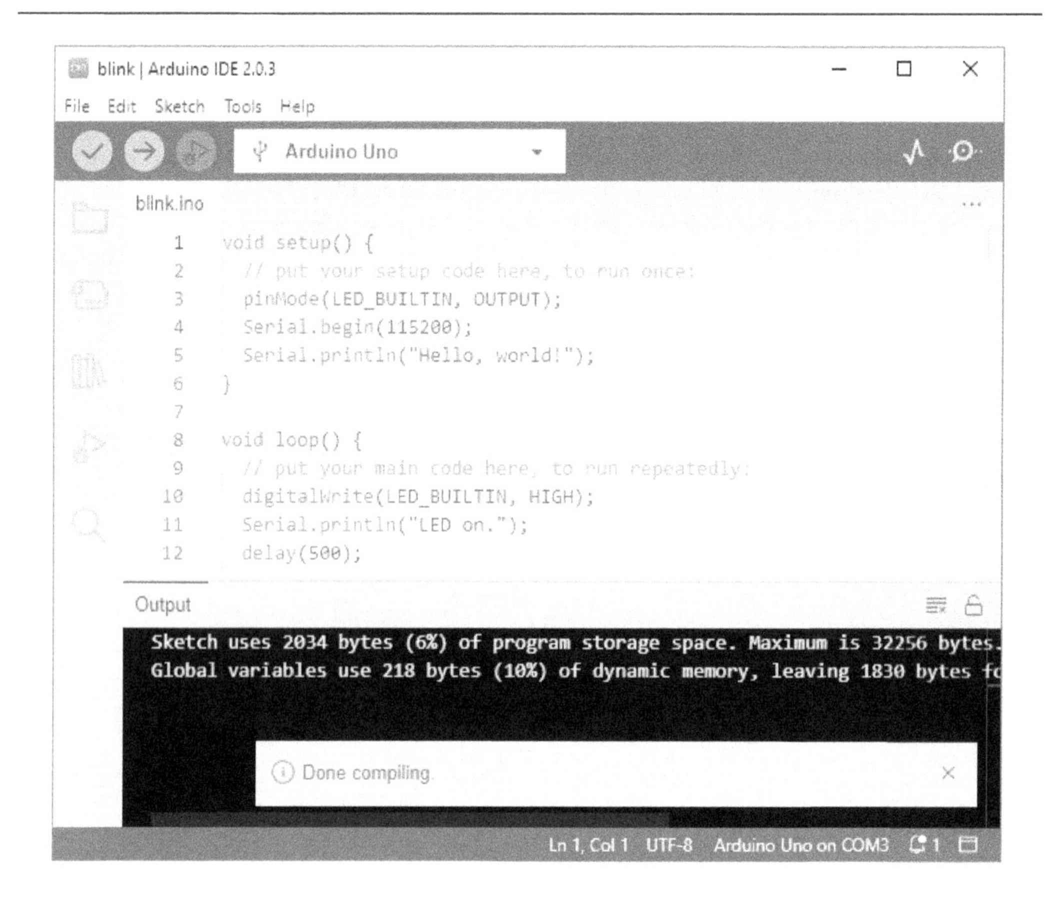


Figure 1.7 – How the IDE looks after successful compilation

The information in the console can be useful to keep an eye on for larger programs. If any of these values get close to 100%, the program can become unstable during execution. In this case, you will want to either optimize your program to use less storage and dynamic memory or upgrade to a more capable Arduino board that has more of them.

## Transmitting the code to your Arduino board and running it

If everything is set up correctly, it is time to transmit your code to the Arduino! To start this process, click the arrow symbol in the top-left corner of the IDE. This will trigger the familiar compilation process, and as soon as the IDE is done compiling, it will start transferring the compiled machine code to the Arduino board via the USB-to-serial programmer interface. This should only take a short amount of time and is accompanied by two new pop-up messages: **Uploading...** and, if all goes well, **Done uploading**. If you look at the TX and RX LEDs on your Arduino board, you will see both blinking very rapidly during the transmission.

After the program is successfully transmitted, the Arduino board will immediately start running it. If all went well, you should see the status LED turn on and off at a frequency of 1 Hz. But we also put some serial print commands in our code. To see the content of these messages, we can use the serial monitor that is built into the Arduino IDE. In the IDE, click on the magnifier glass icon in the top-right corner to open the serial monitor. It will appear as a new tab under the editor area, next to the console. You will need to select the baud rate from the drop-down list in the right corner of the serial monitor that matches the one in the `Serial.begin()` command of our sketch – **115200** in our case. As soon as you set the correct baud rate, you should see alternating lines of `"LED on."` and `"LED off."` appear. And if you push the *RESET* button on your Arduino to start the program from the beginning, you will see the line `"Hello, world!"` appear once. This shows you that the `setup` function is indeed executed once at the beginning of the program. *Figure 1.8* shows the output of the serial monitor, as well as the baud rate setting:

Figure 1.8 – Expected output in the serial monitor

You will also see the TX LED flicker every time the status LED switches on or off. This is signaling that the Arduino is transmitting serial data, namely the messages (`"LED on."` and `"LED off."`). You can use the serial monitor to send a message to your Arduino board. To do this, type anything you like into the line at the top of the serial monitor and press *Enter*. This will send your text, character by character, to the Arduino board. The moment you press *Enter*, you will see the RX LED light up for a brief moment, indicating that data is in fact transmitted from the PC to the main microcontroller. We did not program our Arduino to react to serial data, so it will just ignore it and it has no effect on the execution of our program. But with a few minor changes to our program, we could control the frequency of the blinking with a message from the serial monitor, for example.

Congratulations, you have successfully written your first Arduino program and it is running right now on an Arduino Uno board on your desk! Now that everything is set up and working, you might want to play around with the program a little bit to get more familiar with the workflow. You can change the words that are being printed or the frequency of the blinking, or you can try to change the serial baud rate (both on the Arduino and on the PC side). You can also see what happens when there is an error in your program code – for example, when you delete a semicolon at the end of a line. In this case, the compiler will fail, but it will give you a useful error message and even highlight the area of code where the problem lies.

# Examples of Arduino robot technology

Hopefully, you are excited about the fact that you are now able to program your Arduino microcontroller board. You can be sure that this is only the very beginning of our journey toward building real robots with our newly learned Arduino skills. Arduino technology is often used for research and development to quickly prototype hardware concepts, or even to build cost-effective production systems. Researchers and engineers use Arduino boards on every kind of robot, from solar-powered autonomous boats to monitor the water quality of lakes to scientific data collection on rockets and high-altitude balloons.

There are highly active open source software projects that let you turn your Arduino into a flight controller for autonomous airplanes (`https://ardupilot.org/`) or into the brains of your DIY 3D printer (`https://marlinfw.org/`). You just took the first step toward being able to do all these things yourself and become part of the active community of hobbyists, enthusiasts, researchers, and engineers who use Arduino for their robotics projects.

# Summary

In this chapter, we have introduced the Arduino ecosystem with its two main components, the Arduino IDE and Arduino boards. We have taken an in-depth look at the versatile Arduino Uno and you have transferred your first program onto your own Arduino Uno, using the entire Arduino toolchain. We will build on all of this in the next chapter, where we will learn how to use the Uno's interfaces to communicate with sensors, make motors spin, and control entire robots.

# Further reading

The Arduino IDE and the Arduino Uno board are very well documented. Here are a few resources that can help you learn more about them:

- The landing page of the well-maintained Arduino documentation, from which you can dive deeper into just about any topic related to using Arduino: `https://docs.arduino.cc/`
- Some great information about the main features of the Arduino IDE 2.0: `https://docs.arduino.cc/software/ide-v2`
- The official and always up-to-date Arduino guide landing page with links to helpful resources for getting started: `https://www.arduino.cc/en/Guide`
- A great Arduino getting-started guide that touches upon many topics that we will cover in more detail in the following chapters: `https://docs.arduino.cc/learn/starting-guide/getting-started-arduino`
- The official documentation of the Arduino Uno R3 board: `https://docs.arduino.cc/hardware/uno-rev3`
- A very detailed overview of the functionality and capability of every pin on the Arduino Uno: `https://commons.wikimedia.org/wiki/File:Pinout_of_ARDUINO_Board_and_ATMega328PU.svg`

# 2

# Making Robots Perceive the World with Sensors

Every robot needs sensors to perceive its environment, and sometimes to monitor itself. This chapter introduces some essential background of sensor technology in general and presents specific sensors that are the most useful, most accessible, and easiest to integrate into an Arduino-based robot. We will take a look at how your Arduino can communicate with these different sensor types, and we will go through code examples that show how to use the most common interfaces (GPIO, analog, I2C, **Serial Peripheral Interface** (**SPI**), and Serial).

This chapter is divided into the following sections:

- Understanding common sensor characteristics
- Commonly used sensor types for Arduino robots
- Common sensor interfaces and code examples

## Technical requirements

We will use a variety of sensors in this chapter to exemplify the various interfaces. You do not need to buy each of these sensors to get the most out of this chapter. However, it will be helpful if you reproduce at least the potentiometer example. For this, you will need your Arduino Uno again, an analog potentiometer (rotary or linear), and three cables to connect the two.

## Understanding common sensor characteristics

The task of a sensor is to turn a feature of the physical world that our robot cares about, for example, the ambient temperature, into an electrical signal that the Arduino can read via any of the interfaces that we will discuss later in this chapter. To select a suitable sensor, we need to be sure that it can at least theoretically meet the requirements of our application. There are many characteristics for any given sensor, but in practice, it can be hard to know at the start of a DIY robotics project what exact requirements really matter. It can sometimes be difficult to find the exact technical details of a sensor.

That being said, the two characteristics that you should always consider when selecting a sensor for your robots are the sensor's measurement range and resolution.

The **measurement range** tells us what the minimum and maximum values of the physical quantity that our sensor measures are. A distance sensor, for example, might have a minimum range of 10 cm and a maximum range of 100 cm. If the actual distance is outside this range, the sensor might report an error code or, worse, erroneous values. If our project is to build a robot that needs to follow another robot at a distance of 1 cm, a distance sensor with the above measurement range would a bad choice, no matter all its other characteristics. However, if we are building a drone that needs to automatically hover at a height of 50 cm, this sensor might just be perfect because the target height is almost at the center of the range, and we do not expect the drone to deviate from this value by more than a few centimeters. Before selecting a specific sensor, we need to consider what range we need to measure (of distance, temperature, G-force, brightness, etc.) and select a sensor that covers this entire range, ideally with some margin.

The **measurement resolution** tells us, broadly speaking, how precisely the sensor can measure within its measurement range. Higher resolution means more precise, so sensors with higher resolution are generally more desirable. In the example of our distance sensor, its resolution might be 1 cm. This means it takes a change in distance of 1 cm for the sensor to even recognize that something has changed. Or, in other words, as long as the distance changes within 1 cm, the sensor output might not change at all. Whether this is acceptable or not depends on the application. For the drone hovering at a height of 50 cm, sensing the height with 1 cm resolution might be more than enough. If the sensor is used for a robot that uses it to precisely determine the dimensional accuracy of a production process, this sensor would likely not be the right choice.

There are two main ways of expressing a sensor's resolution: **absolute** and **relative**. In our example of a distance sensor with a 1 cm resolution, the resolution is expressed in absolute terms in the unit of measurement. This is the most convenient and easy to work with. Counterintuitively, a smaller value for the absolute resolution means higher resolution, since smaller steps imply that there are more steps in the measurement range. Other examples of absolute resolution are a current sensor with an absolute resolution of 1 mA or a temperature sensor with an absolute resolution of 1 K. The second common way of expressing sensor resolution is as a relative resolution in **bits**. For example, you might see a current sensor with a measurement range from -5 A to 5 A and 10-bit resolution. This means that it can detect, $2^{10} = 1024$ distinct states across its measurement range, and this is equivalent to stating that it has an absolute resolution of the following:

$$\frac{\mathbf{5}\boldsymbol{A} - (-\mathbf{5}\boldsymbol{A})}{\mathbf{2}^{\mathbf{10}}} \approx \mathbf{10}\boldsymbol{mA}$$

To compute the absolute resolution in physical quantities from the relative resolution in bits, we need to know the measurement range. Providing the relative resolution is often useful when the measurement range is not fixed. Many sensors have a configurable measurement range so that they can be tuned for a specific application. We can, for example, reduce the range to get an increased absolute resolution or increase the range at the cost of a lower absolute resolution. This is also practical since it tells us immediately how much data we need to transfer per measurement. A value with 8-bit resolution will fit into a single **byte**, but for 10 bits we will need space across 2 bytes.

It is useful to keep in mind that a sensor's theoretical resolution is not necessarily what we really get in terms of signal quality. The *effective resolution* is typically lower. Almost all sensors suffer from signal noise, which means that the sensed signal jitters up and down randomly to some extent even if the input to the sensor does not change. This part of the signal is usually useless. If the noise is larger than the sensor resolution, then this part of the resolution is wasted on measuring the signal noise and does not actually provide us with more information (unless we do some clever signal processing). Sensors also have other systematic limitations to their performance, such as limited accuracy, sensitivity, and repeatability, that limit the signal quality even if their resolution increases. In practice, it is often not useful to choose a sensor with a higher resolution than necessary. When we use very high-resolution sensors, we often need to spend significant effort on things such as eliminating signal noise to provide them with a good enough input to make use of their high resolution. A good example of this are infrared cameras on space telescopes. These are typically cooled to extremely low temperatures with complex coolers to limit signal noise that would be picked up by their very high-resolution and very sensitive imagers. Sometimes it might also make sense to artificially lower the resolution (sometimes called downsampling). For example, when we use a 10-bit torque sensor but the signal is so noisy that the effective resolution is really only 8 bits, we can simply ignore the two least significant bits. In this case, we only need to transmit 8 bits without losing any (useful) information, potentially simplifying and speeding up data transmission.

There are many other characteristics for any given sensor, but none of them are as important for DIY robot projects as range and resolution. The following list gives you an overview:

- **Linearity**: Many algorithms assume a linear correlation between the input to the sensor (for example, the distance to a wall) and the output of that sensor. So, if the distance doubles, the sensor output is also assumed to double. Most sensors are in fact relatively *linear* across their measurement range. However, some sensors are not linear at all. If we know the mathematical description of the nonlinearity, we can write code to linearize the signal, but that usually has undesirable side effects such as a variable resolution across the measurement range.
- **Hysteresis**: This means that a sensor will output a (slightly) different measured value for the exact same input, depending on whether the input is currently increasing or decreasing. In practice, most commercial sensors have a small enough hysteresis that we can safely ignore it in DIY projects. But if accuracy matters a lot, it is good to be aware of this phenomenon.
- **Reaction time**: Generally speaking, the change in a sensor's output is always at least a little bit delayed with respect to a change in the physical property it is measuring. In most DIY robotics applications, we do not have to worry too much about this delay. One common exception to this can be temperature. Depending on the type of temperature sensor and especially the way it is attached to our robot, there might be a significant delay between the actual temperature change (inside a motor for example) and the temperature value measured by the sensor. This can be problematic if we want to turn off a motor before it overheats. It is, therefore, good practice to mount temperature sensors as close as possible to the part we are concerned about and ensure good thermal coupling. This will minimize the delay of the sensed temperature signal.

- **Update rate**: This quantity, also called sampling rate, is related to reaction time. It tells us how fast we can acquire new samples from a sensor and is typically expressed in Hz (samples per second). The fastest we need to sample sensors in Arduino robots is typically on the order of 100 Hz, and a sensor with an update rate of more than 1 kHz is usually not necessary (there will likely be other bottlenecks in our program preventing us from taking advantage of that high of an update rate anyways). However, if the update rate is too slow, it could negatively affect our robot's performance.

With this quick overview of common sensor characteristics out of the way, let's look at the types of sensors that lend themselves particularly well to Arduino robotics projects.

# Commonly used sensor types for Arduino robots

There are so many distinct types of robots that it is impossible to make a general recommendation for the right sensors to choose for your project. But it is useful to know what sensors are out there so you know what you can choose from. We will look at a list of sensor types in this section, and then revisit some of them later in the chapter as examples for several types of interfaces. Most of the parts that we will call sensors are technically **sensor modules**. The actual sensor is only one part of it, and there are additional parts such as voltage regulators, signal conditioning circuitry, and **Analog-to-Digital Converters** (**ADCs**). It is not uncommon for all these parts to be integrated into a single chip.

## Switches and buttons

Arguably the simplest sensor types are switches and buttons that either open or close a contact when pressed, and do the opposite when released. Simple as they are, they might be all that your first mobile robot needs to detect contact with walls and obstacles and change direction accordingly. Switches and buttons are ubiquitous, on appliances, gadgets, cars, planes, TV remotes, and so on. While they come in all kinds of shapes and sizes, their operating principle is always the same: they open or close an electric circuit. It is extremely easy to read the state of a switch with an Arduino. One thing to be aware of when reading a button at a very high update rate is *bouncing*, a phenomenon that many types of buttons produce. During a very short time span when the button makes or breaks internal contact, it opens and closes several times. This is like a ball falling on the ground – before it settles, it will bounce, making and breaking contact with the floor several times. Button bouncing is good to be aware of during debugging when your robot behaves unexpectedly, and there are straightforward ways to *debounce* a button in software, if needed.

## Distance sensors

Especially for mobile robots, distance sensors are often used to detect walls, drops, or other obstacles. If you have ever backed up in a car that beeps increasingly intensely the closer you get to the wall of the garage, it was most likely using several dedicated distance sensors. Some cars also use distance sensors to keep a minimum distance from cars in front of them. Autonomous cars often use several, very sophisticated distance sensors to model their surroundings in three dimensions. Distance sensors for Arduino robots include the following:

- **Ultrasonic distance sensors**: These work with sound that is inaudible to the human ears. They consist of a pair of two devices on a common PCB, an ultrasonic emitter (think: speaker) and a receiver (think: microphone). They often have the necessary electronics integrated to generate the ultrasonic waveform to drive the emitter and convert the output of the receiver into a distance value. They are cheap and robust, but they are also relatively large. Their update rate is low and their measured value is the average over their large field of view. They are good for sensing the distance to walls or the palm of your hand, but not for detecting anything smaller than that. One key advantage of these sensors compared to light-based sensors is that they work in foggy or dusty environments.
- **Infrared distance sensors**: These sensors emit a focused beam of invisible light with an infrared LED. They determine the distance to an object by analyzing the direction of the reflection with an infrared detector (see *Figure 2.1*). The emitter and the detector are usually housed in the same enclosure. These sensors are cheap and fairly precise at small distances, with a narrow field of view. However, they tend to be highly nonlinear and require a little bit of extra code to linearize their output signal.

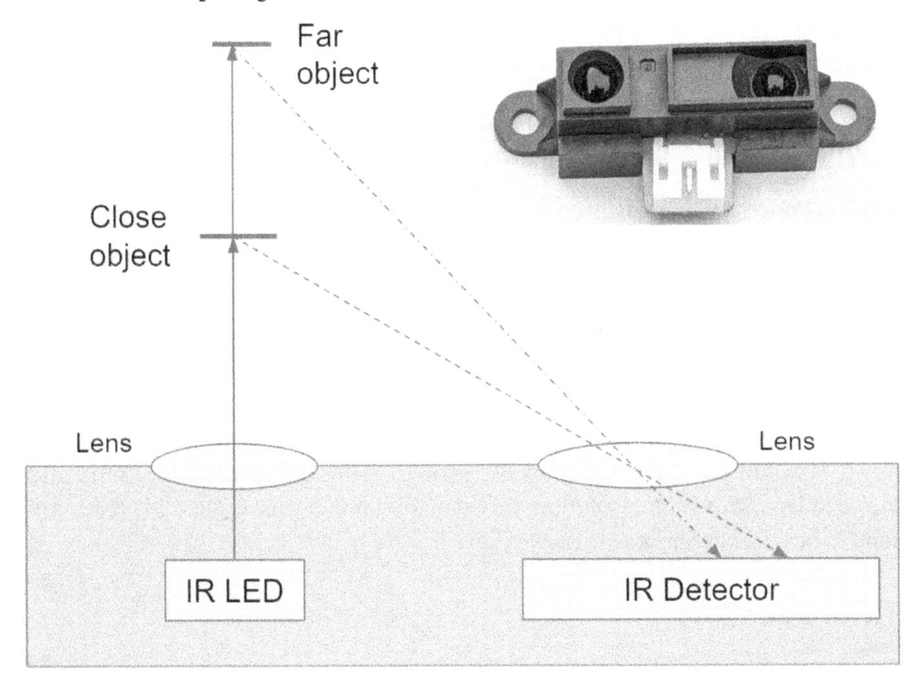


Figure 2.1 – The principle of operation of infrared distance sensors

- **Time-of-flight (ToF) distance sensors**: These sensors also consist of a light emitter and detector. They measure the time it takes the light from the emitter to reach the object in front of it and then reflect back to the detector, hence the name. These sensors are the top choice for precise measurement with a very narrow field of view and thus great spatial accuracy. When the light that is used is laser light, they are called **Light Detection and Ranging** (**LiDAR**) sensors. ToF sensors are usually the most expensive option and often the most difficult to use due to relatively complex interfaces and a variety of settings. But their performance tends to be superior to the other options.

Which of these distance sensors is best for your robot depends on the requirements, primarily, the range that the sensor needs to cover as well as the required resolution and accuracy. If you need to measure the distance that your robot has traveled, you can also use encoders on the wheels.

## Encoders (rotary and linear)

Encoders sense their current position along a predefined trajectory, either a rotation or a linear path. Rotary encoders are often attached to electric motors to keep track of the position of the motor shaft, or to the wheels of a mobile robot to keep track of the wheel's rotations. Linear encoders are less commonly used but often work in the same way as their rotary counterparts, and you can attach them to a pneumatic piston for example to sense its extension. Encoders can be divided into two main categories:

- **Absolute encoders** know their absolute position along their trajectory the instant they are powered on. Potentiometers are the most commonly used type of absolute encoder.
- **Relative encoders** incrementally count steps along their trajectory (for example, black/white transitions along an optical track), and the sum of transitions translates to their current position. However, this position is *relative to the position where they were turned on* and started counting. Incremental encoders are mostly optical or magnetic (counting white/black or north pole/south pole transitions). Relative encoders usually output two pulsed signals in quadrature encoding that tell the Arduino when a step occurred, and in which direction it went (positive or negative).

Generally speaking, it is easier to make and integrate relative encoders into your robot than absolute ones. However, it will often be the Arduino CPU's job to count the sensed transitions, and for a fast-moving encoder with a high resolution, this can add up to a significant CPU load. Absolute encoders tend to be larger and more expensive.

## Temperature sensors

Temperature sensors are usually small and simple sensors with either an analog or digital interface that measures their own temperature. To make them sense the temperature of something relevant, such as a motor, we have to make sure they are thermally well coupled to that motor. This means ensuring they have good, direct contact, and ideally adding some thermal paste in between them.

## Pressure sensors

Pressure sensors can measure the pressure in a gas or liquid. The most common application in robots is the barometric pressure sensor, which measures the pressure of the ambient air relative to a (tiny) gas cavity inside of them. The same sensor type is often found in watches designed for outdoor activities. They are extremely useful for flying robots to estimate the height over the ground since ambient pressure decreases with height. Similarly, underwater robots use pressure sensors to sense their diving depth. When configured as differential pressure sensors and paired with a **pitot tube**, pressure sensors can also be used to sense air speed in flying robots.

## Light sensors

Light sensors sense the amount of light that shines on them. High-quality light sensors that enable repeatable, quantitative measurements can be expensive and are usually not needed for robotic applications. The type of light sensors used in Arduino robots give rather qualitative measurements, but this is often enough. Light sensors can be used to build light barriers, for example, to detect whether something is between the jaws of a gripper. Small light barriers paired with a spoked wheel can be used to build a relative encoder (the light barrier will be interrupted each time a spoke passes through it). Light sensors inside a box can be used to detect whether the lid is open or closed, to program a robot to only move in the dark, to help solar panels track the sun, or to detect the position of a black line on the ground. The simplest light sensors are photo resistors that decrease their resistance as more light hits them. There are also phototransistors that work similarly, increasing their conductivity as they are exposed to brighter light. Light sensors are very cheap, small, and easy to use, so it is always worth considering whether a given sensing problem can be solved with them. There are advanced sensor modules that can sense the color of incoming light (or of a reflective surface if they have active lighting). These can be very useful if you want to build a robot that sorts things by color but are a little more complicated to use than a simple light sensor.

## Sound sensors

Microphones can be used to sense sound. However, it is relatively difficult for an Arduino board to properly sample a microphone since this task typically requires high-resolution analog-to-digital conversions at very high sample rates. Fortunately, there are integrated sensor modules available that pair a microphone with dedicated electronics that analyze the microphone's output and generate a digital signal that simply indicates whether the sound level is above or below a set threshold. This can be very useful, for example, to make a robot react to your clapping.

## GPS

A **Global Positioning System** (**GPS**) receiver can be used to determine a robot's location anywhere on earth to within a few meters. To this end, the receiver has to detect signals from several GPS satellites simultaneously. While a GPS receiver is a small, inexpensive, and relatively simple device, the components of the GPS infrastructure (and its counterparts such as GLONASS, Beidou, and Galileo) are incredible technological masterpieces. Because the GPS signal is weak, a GPS receiver requires a direct line of sight into the sky and usually does not work well indoors. GPS positioning can be a very useful addition for outdoor robots that cover large distances, such as automatic lawnmowers and flying or swimming robots. In practice, you will want to pair the receiver with an external GPS antenna (which is usually bigger and heavier than the receiver itself) for good performance; see *Figure 2.2*.

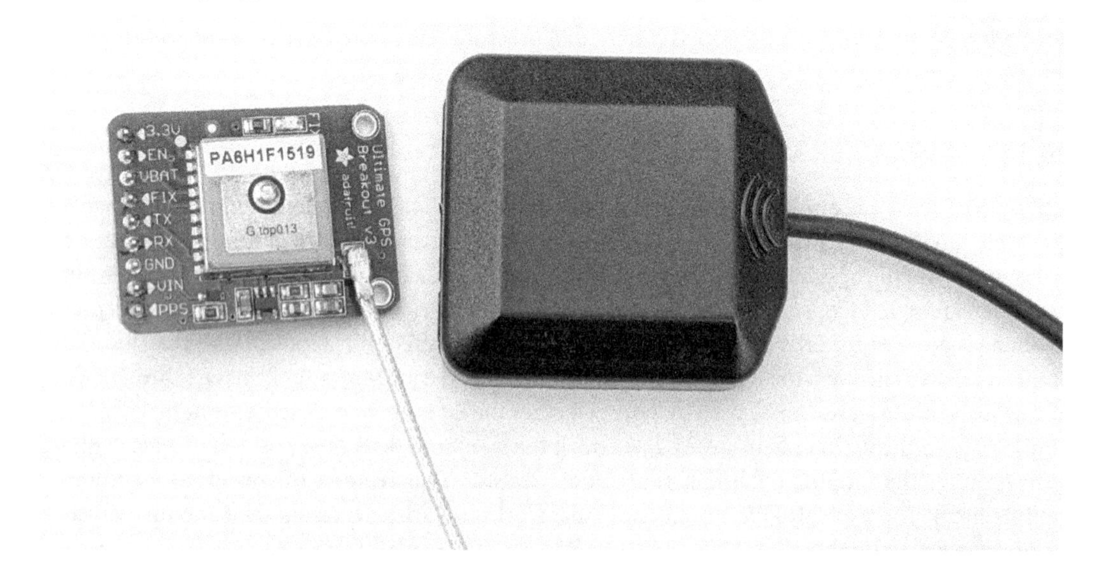


Figure 2.2 – A GPS receiver (left) with its external antenna (right)

There are special types of GPS receivers with **Real-Time Kinematics** (**RTK**) functionality that achieve positional accuracy on the order of 10 cm. However, these are significantly more expensive.

## IMU

An **Inertial Measurement Unit** (**IMU**) is a combination of at least two types of inertial sensors: **accelerometers** and **gyroscopes**. Accelerometers measure their linear acceleration, typically in three dimensions. Gyroscopes measure their absolute rate of rotation, also in three dimensions. The most common application for IMUs is to estimate the three-dimensional orientation of the robot, for example, expressed in yaw, pitch, and roll angles. To avoid measurement drift around the yaw axis, IMUs are often paired with magnetometers that sense the earth's magnetic field in three dimensions as a steady reference. IMUs that integrate magnetometers are sometimes called **9DoF IMUs**. Computing the 3D orientation from the nine measured values requires sophisticated sensor fusion. The necessary computations can be strenuous for an Arduino microcontroller and, if possible, it is preferable to use an IMU that does the sensor fusion on board.

IMUs are crucial sensors for stabilizing autonomous airplanes and quadrotor drones, as well as two-wheeled self-balancing or walking robots. Accelerometers and gyroscopes can also be used individually, for example, for logging the acceleration of a rocket during flight or controlling the spin rate of a rotating LiDAR without the need for an encoder. Inertial sensors used in DIY Arduino projects are usually based on **Microelectromechanical Systems** (**MEMS**) technology, which makes them small, robust, and cheap.

After this bird's-eye overview of a lot of different sensor technologies and their application, let's get into the weeds and learn how we can actually use them with Arduino and see some of them in action.

# Common sensor interfaces and code examples

Now that we have gotten to know a few types of very useful sensors for Arduino robots, we need to understand how to *talk* to them. As we mentioned in *Chapter 1, Introducing Robotics and the Arduino Ecosystem*, an Arduino board such as the Arduino Uno has many different types of interfaces and the Arduino IDE provides us with the software building blocks to use these interfaces very easily. This section will walk you through the most relevant interfaces and show example code for each of them that communicates with a specific sensor.

## GPIO – great for simple sensors

**General-Purpose Input/Output** (**GPIO**) includes all the digital pins of our Arduino. For the purpose of this section, we will only use its input functionality. To interface with sensors, it can be used in two different modes (with and without an internal pull-up resistor), and some GPIOs can serve as external interrupt triggers to capture inputs with very high timing accuracy.

### *Digital inputs*

A simple button or switch has two contacts, and they are either connected when the switch is closed or not connected when the switch is open. To read the state of the switch (open or closed), we simply need to connect one of its terminals to one of the Arduino's digital pins, and the other one to GND. In the `setup()` function of our program, we configure the pin as a digital input and enable its internal pull-up resistor. The pull-up resistor will make sure that the pin is reliably pulled to a logic HIGH level when the switch is open. When the switch is closed, it connects the pin directly to GND and pulls it to a LOW level. This makes this switch a so-called *active low* switch. Using the `digitalRead()` function, we can read the pin's logic level and thus the state of the switch. *Figure 2.3* shows the wiring for an active low push button switch connected to pin 2 of an Arduino Uno.

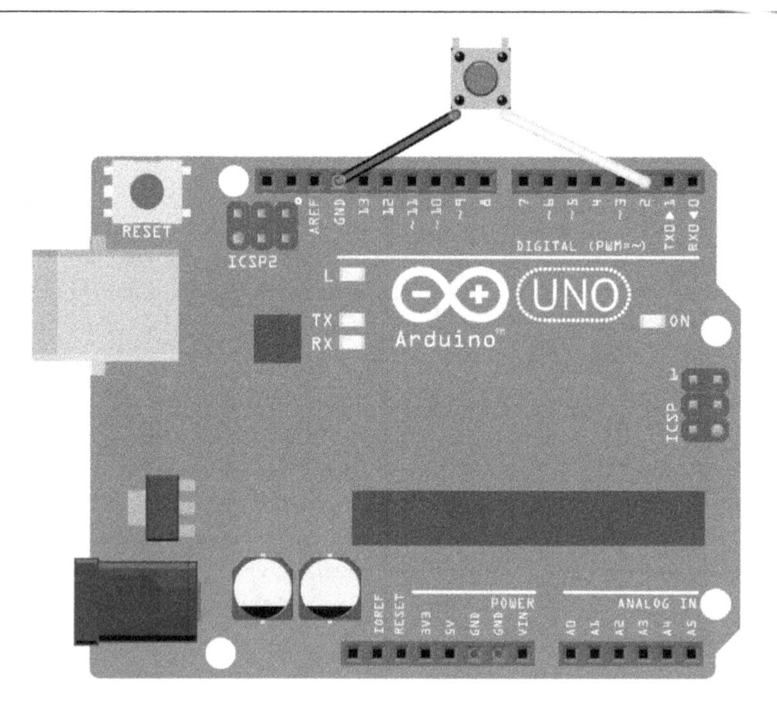


Figure 2.3 – Wiring of a simple push button as active low switch on pin 2

For robustness, it is usually preferable to use external pull-up resistors over the ones internal to the microcontroller, but it is by no means necessary. If you want to invert the signal of your switch to make it an *active high* switch, you can connect an external pull-down resistor to the input pin (the Arduino Uno does not have built-in pull-down resistors) and connect the other terminal of the switch to 5V instead of GND.

### Interrupt pins

Instead of reading the state of a digital input pin with the `digitalRead()` function, some pins on Arduino boards can be configured as **interrupt** pins and automatically react to a change in the pin's voltage level. If the state of an interrupt pin changes, the current program execution is interrupted (hence the name) and a function that we *attached* to this interrupt source, called the **Interrupt Service Routine** (**ISR**) is called to *service* this interrupt. This makes it easy to react efficiently to fast-changing signals. Common use cases for hardware interrupts are decoding **Pulse Width Modulation** (**PWM**) signals from RC receivers or the pulses generated by an incremental encoder. We will encounter these use cases in a later chapter. But for now, we can also use an interrupt mechanism to sample a simple button or switch. It is good practice to keep ISRs as short as possible and to declare variables that are shared between the ISR and the main program as `volatile` to avoid unintended side effects.

The following example code uses an active low switch on the Arduino's pin 2 to control the onboard LED. When the button is pressed, it lights up, and when released, it turns off. The wiring of the switch is the same as shown in *Figure 2.3*.

The following code exemplifies both methods of sampling a digital input pin, that is, via polling and via an interrupt. In the `loop()` function, the button is polled with `digitalRead()` and the LED is turned on or off accordingly. We set the value of the LED pin to the inverted value of the switch state since it is an active low switch, but we want the LED to turn on when the button is pressed. Additionally, we attach an ISR to every change of the pin state and let it increment a variable to count the state changes. In the main loop, the value of this counter is printed to the Serial Monitor every 100 ms. Unless you use an extremely high-quality button, you will see the counter increment by (much) more than one with every button press and release. This is caused by the bouncing effect and the capability of the interrupt mechanism to capture changes that happen very fast. While the bouncing does not matter if all we want to control is the LED, if our program relied on an accurate count of the button events then we would need to add some code for *debouncing*.

To achieve periodic printing without the use of `delay()`, we use the `millis()` utility function. It returns the time in milliseconds since the Arduino was turned on. The modulo 100 of this value will be exactly zero every 100 ms, and we use this fact to periodically trigger the call to `Serial.print()`. This is a simple but not great way to trigger periodic events. We will later get to know a much better method:

```
// We declare pin numbers as constants.
const int button_pin = 2;
// We declare ISR variables volatile.
volatile unsigned long num_button_events = 0;

// Interrupt service routine.
void buttonIsr() {
  num_button_events = num_button_events + 1;
}

void setup() {
  // Start the Serial interface.
  Serial.begin(115200);
  // Set LED pin as output.
  pinMode(LED_BUILTIN, OUTPUT);
  // Set button pin as input. Activate pull-up.
```

```
  pinMode(button_pin, INPUT_PULLUP);
  // Attach the ISR to the interrupt source:
  // Any change of the state of button_pin.
  attachInterrupt(digitalPinToInterrupt(button_pin), buttonIsr,
CHANGE);
}

void loop() {
  // Read the button state (HIGH or LOW).
  bool button_state = digitalRead(button_pin);
  // Control the LED accordingly.
  // Invert input to turn LED ON when input is LOW.
  digitalWrite(LED_BUILTIN, !button_state);
  // Periodically print number of state changes.
  if (millis() % 100 == 0) {
    Serial.println(num_button_events);
  }
}
```

If you do not have a button or switch on hand to try this out, you can simply use a jumper wire: plug one end into one of the Arduino's GND terminals and connect the other one to pin 2 to emulate a button *press* and unplug it to emulate a button *release*.

## Analog is simple

All analog inputs of an Arduino are sampled by a single built-in 10-bit ADC. In its default configuration, the reference voltage is 5V, which means that the measurement range is between GND and 5V. We can use a different internal reference or provide a reference voltage via the AREF pin to select a smaller range and increase the absolute resolution of the measurement, but for now, we will just work with the default 5V reference. If we connect the analog input to GND, the `analogRead()` function will return `0`. If we connect it to 5V, it will return `1023`. A voltage in between will result in a return value between these two extremes. To read an analog sensor, we simply need to provide it with power, connect its analog output to an Arduino analog input, and use `analogRead()` to read a value between `0` and `1023`. We can easily convert this value back to a voltage reading, but this is often not even necessary.

The analog interface is beautifully simple both in hardware and software. It only needs one wire for the signal and only one line of code to read it. Its biggest downside, however, is the limited resolution at a maximum of 10 bits. In practice, there will always be signal noise that reduces the effective resolution even further. Signal noise typically increases with wire length and the presence of sources of **Electromagnetic Interference** (**EMI**) such as motors or voltage converters. Signal noise can cause difficult-to-debug issues with analog signals. Consequently, it is best to keep analog signal cables as short as possible and away from EMI sources or use shielded cables, if possible. Other methods of reducing signal noise are analog or digital signal filtering, but these filters will inevitably introduce undesired signal delay. Even though analog sensors are very appealing due to their ease of use, it is generally preferable to choose a sensor with a digital interface whenever possible and avoid all the potential problems that analog signals can bring with them.

The simplest and most widely used analog sensor for Arduino projects is the **potentiometer**. It has three terminals and they effectively implement a resistive voltage divider. When the two outer terminals are connected to GND and 5V, the voltage between the center terminal and GND will be somewhere between 0V and 5V, depending on the position of the knob. We can read this voltage with an analog input pin of the Arduino, which allows us to sense the position of the knob. The schematic in *Figure 2.4* shows how to connect a potentiometer to the Arduino Uno. The resistance value of the potentiometer is not important, as long as it lies somewhere in the range of 1k Ohms – 200k Ohms.

Figure 2.4 – Connecting a potentiometer to the Arduino Uno

The following code periodically samples the output voltage of an analog sensor such as a potentiometer connected to pin A0 and prints it to the Serial terminal. The program also prints the minimum and maximum values of `0` and `1023`. This is useful to avoid the Serial Plotter of the Arduino IDE from autoscaling the plot:

```
// Use pin A0 for analog input.
const int analog_pin = A0;

void setup() {
  // Start the Serial interface.
  Serial.begin(115200);
}

void loop() {
  // Read the potentiometer voltage.
  int analog_value = analogRead(analog_pin);
  // Print analog value every 50 ms.
  if (millis() % 50 == 0) {
    // Print analog value.
    Serial.print(analog_value);
    // Tab character creates a new line in the plotter.
    Serial.print('\t');
    // Print min and max range to avoid auto-scaling.
    Serial.print(0);
    Serial.print('\t');
    Serial.println(1023);
  }
}
```

If you open the Arduino IDE's Serial Plotter, you will see three lines: one horizontal line at `0`, another one at `1023`, and a live plot of the sensor output in between, as shown in *Figure 2.5*.

Figure 2.5 – The resulting plot of turning a potentiometer back and forth

In this screenshot of the Serial Plotter, the plot is stopped with the **STOP/RUN** toggle button in the top-right corner. While stopped, you can hover with the mouse over the lines to see the numeric values at any position. You can also uncheck the boxes for **value 2** and **value 3** (the min and max values) to enable auto-scaling.

Even though the analog interface is amazingly easy to use, it has many severe limitations that make it often not very practical. That is when digital serial interfaces come into play.

## Serial interfaces

Most sensor types require a little more communication than the analog interface can provide. Many sensors have various settings that need to be written before we can use them, and their output resolution might well exceed the 10 bits that the Arduino's ADC can resolve. In this case, they need a more powerful interface type, and that is where digital serial interfaces come in. Whereas with an analog interface all the data is transmitted at once as the voltage of an analog signal, a serial interface transmits a sequence of tiny bits of information (literally called bits) one after the other. All these bits together form a binary message that contains the sensor signal and any other information that we want to transmit. There are different protocols that set different standards for hardware and software to transmit data serially, each optimized for different performance criteria such as speed, low protocol overhead, minimum hardware requirements, or ease of implementation in software. The most common serial interfaces used in Arduino projects are **Inter-Integrated Circuit** (**I2C**), SPI, and Serial. If multiple ICs can communicate over the same physical interface, this interface is called a bus.

### *I2C is great for multiple sensors*

I2C is a simple serial interface developed to connect multiple ICs on the same circuit board, or multiple circuit boards over short wires. Because multiple ICs can be part of the same I2C setup, I2C is considered a communication bus. This interface is primarily optimized for hardware simplicity and only needs two wires to work, one for a shared *clock* signal and one for *data*. An I2C bus has two distinct kinds of members, a **master** that controls the communication and potentially many **slaves** that can communicate with the master.

> **Important note**
>
> The technical terms master and slave are outdated but deeply rooted in technical documents and standards. To avoid confusion, especially for novices to the topic of data buses, they are mentioned here once. Going forwared, the adoption of more modern and more inclusive alternatives is encouraged. While there is no universal consensus on the best way to replace these words in the technical context yet, controller/target and leader/follower are sensible alternatives. The official Arduino documentation uses controller/target, and the remainder of this book will follow suit.

The Arduino board is the controller and all attached sensors are the targets. The controller generates a clock signal on the clock line (**Serial Clock** (**SCL**) and initiates data transmissions on the data line (**Serial Data** (**SDA**). To send a bit over either of the two lines, the controller and the targets can connect the line to GND (logical LOW). For the lines to go back to 5V (logical HIGH) once released, they rely on external pull-up resistors that we need to provide. *Forgetting to add pull-up resistors to an I2C bus is an easy mistake to make and is a surefire way to create I2C problems.*

The SDA pins of the controller and all targets are wired in parallel, and the same is true for SCL. To allow the controller to talk to a specific target (as opposed to talking to all at once), each target needs a unique address. Standard I2C supports up to 128 unique addresses per bus. However, in practice, many I2C targets have either a fixed address or only very few to choose from. *If two targets on the bus have the same address, the functionality of the bus is compromised.*

Because of its simplistic wiring, even with many targets on the same bus (provided that all have unique addresses), I2C is a very popular bus for Arduino projects. The main downside is that data transfer is relatively slow, both due to hardware limitations but also due to protocol overhead. Part of this overhead is transmitting the target address at the beginning of every transaction. While I2C allows for bi-directional communication (duplex), data on the single SDA line can only be transmitted in one direction at a time, either from the controller to the target or from the target to the controller. This characteristic makes it a **half-duplex** interface and contributes to its relatively low performance.

The Arduino IDE provides us with the **Wire** library, which makes using the I2C peripheral of Arduino boards extremely easy. To use a library in our Arduino code, we must *include* it with the `#include` statement. On the Arduino Uno, the SDA pin is A4 and the SCL pin is A5. We cannot use these pins as analog input pins in applications that use the I2C interface. Be sure to add external pull-up resistors to these pins when using them as I2C pins. 4.7k Ohm is usually a good value.

In this example, we periodically read and print the temperature from a BNO055 IMU via I2C (it is common for advanced sensors such as IMUs to also include a temperature sensor). The BNO055 breakout board shown in *Figure 2.6* is from Adafruit (`https://www.adafruit.com/product/4646`).

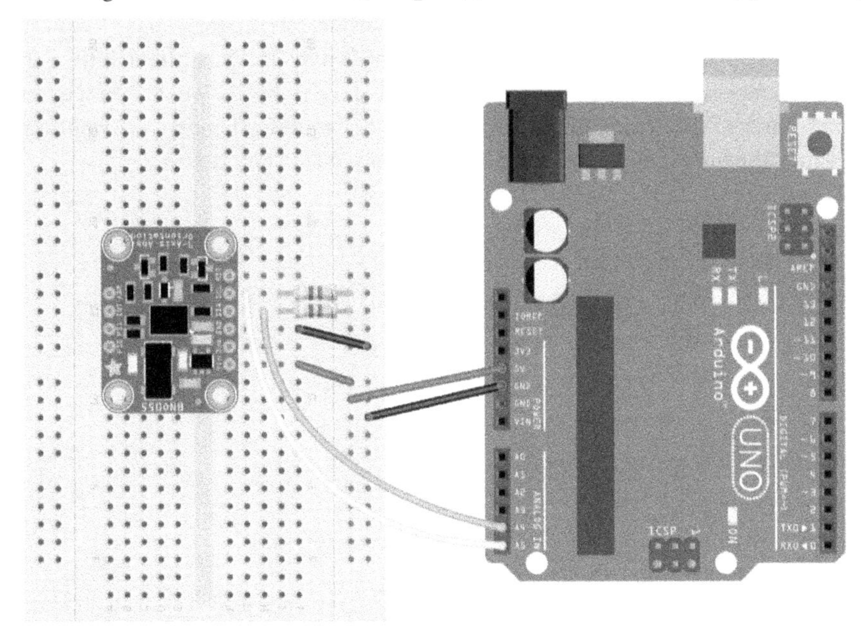


Figure 2.6 – Arduino Uno and BNO055 connected via I2C

In the following code, after including the `Wire` library at the top of the program, we can start the I2C interface with the `Wire.begin()` function inside the `setup()` function:

```
// Include I2C library.
#include <Wire.h>
// I2C parameters of BNO055 chip.
const uint8_t bno_address = 0x28;
const uint8_t bno_temp_register = 0x34;
const uint8_t bno_temp_data_length = 1;

void setup() {
```

```
  // Start the Serial interface.
  Serial.begin(115200);
  // Start the I2C bus.
  Wire.begin();
  // Give the BNO055 some time to start.
  delay(1000);
}
```

The various steps of the I2C transaction are contained in the `bnoReadTemp()` function. The Arduino first begins the transmission with the IMU, identified by its I2C address. It then transmits the address of the temperature register and signals the end of the transmission. It then requests 1 byte of information from the IMU, which will in turn transmit the requested temperature value during the next transmission. The function can then read the transmitted data and return it:

```
uint8_t bnoReadTemp() {
  // Begin I2C transmission with BNO055.
  Wire.beginTransmission(bno_address);
  // Write the address of the temperature register.
  // This requests temperature data from the BNO055.
  Wire.write(bno_temp_register);
  // End the I2C transmission.
  Wire.endTransmission();
  // Request one byte of data from the BNO055.
  // This will be the temperature data.
  Wire.requestFrom(bno_address, bno_temp_data_length);
  // Read and return the requested temperature data.
  return Wire.read();
}
```

The loop function of this sketch simply uses `bnoReadTemp()` to periodically read `BNO055`'s temperature and print it to the Serial Monitor:

```
void loop() {
  // Periodically print the BNO055 temperature.
  if (millis() % 200 == 0) {
    Serial.print("BNO055 temperature [C]: ");
    Serial.println(bnoReadTemp());
  }
}
```

You can see how much overhead (and code) it requires to transmit just a single byte of temperature information, compared to the simplicity of the analog interface. But you can also see in this example how we can communicate in both directions. This allows us to not only request temperature data but data from any of the many data registers of `BNO055` - something that a simple analog interface could not support. The I2C interface scales much easier to larger amounts of data. If the temperature value was 32 bits instead of 8, we would simply request 8 bytes and stuff them into a 32-bit return value (`unsigned long`).

This implementation of `bnoReadTemp()` is meant to exemplify what an I2C transaction looks like under the hood. You will rarely have to write code as detailed as this. There are ready-to-use, open source and free libraries for many sensors already out there that typically expose a much easier interface, even to complex sensors. A good example is Adafruit's library for the `BNO055` module: `https://github.com/adafruit/Adafruit_BNO055`. If we included this library, we could simply use its `getTemp()` function to get the temperature, without worrying about how the I2C transaction looks in detail. The Arduino IDE makes using many of these libraries extremely easy, and we will learn how to work with them in a later chapter.

### *SPI is fast*

**SPI** serves a similar purpose as I2C, but with a stronger focus on speedy data transmission at the cost of more complex wiring. Like I2C, it has a dedicated *clock* line (usually called either SCK, SCLK, or CLK) that carries a clock signal generated by the controller, but it has *two data lines* (**MOSI** and **MISO**) over which data can be transmitted in both directions simultaneously. These acronyms stand for **Master-Out-Slave-In** (**MOSI**) and **Master-In-Slave-Out** (**MISO**), making the data direction clear (even though the terms controller and target are preferred, the terms COTI and CITO are not yet established as replacements for MOSI and MISO). Data can be transmitted on MOSI and MISO simultaneously, triggered by the clock signal. This makes SPI a **full-duplex** bus. Some SPI devices might use different terminology. Microchip, for example, calls the MISO and MOSI pins of the **MCP3008** ADC of our upcoming code example **DIN** and **DOUT** (for data in and data out). This is still unambiguous, since this chip implements an SPI target, making DIN and DOUT equivalent to MOSI and MISO, respectively. SPI devices use **push-pull** line drivers, which means external pull-up resistors are not needed and help increase transmission speeds. Instead of addressing targets by an address that needs to be transmitted, a dedicated active low **chip select** line runs from the controller to each individual target. This reduces protocol overhead and helps avoid the clashing of addresses altogether, but it also means that the number of required wires increases with the number of targets. When fast data transmission is important, for example, to write or read data from an SD card or a camera sensor, SPI is a superior choice compared to I2C.

SPI is a relatively loosely defined standard and allows a broad range of **bitrates**, different **byte orders** (MSB first or LSB first), and different conventions for the polarity of the clock signal and the signal edge that triggers the sampling (called SPI **modes** 0 through 3). You can find the right values for these parameters in the datasheet of the sensor you are using.

Just like with I2C, the Arduino IDE makes it easy to use the built-in SPI hardware of any Arduino board with the **SPI** library. It provides the `SPI.transfer()` function at its core. This function takes 1 byte as an argument and transmits it bit by bit to the target. It does that by cycling the lock line eight times and driving the MOSI line high or low during each clock cycle, depending on the value of the bit to be transmitted. *At the same time*, it will read the state of the MISO line that is driven by the target during each clock cycle and return the 8 received bits as 1 byte. If we are only sending data and are not expecting any useful data from the target, we can just ignore the returned value. The SPI library will generally not handle the chip select pin for us. We must take care of selecting the correct target by pulling its chip select line low ourselves, using a digital output pin and the `digitalWrite()` function.

The following example code uses SPI to communicate with a simple SPI target, the MCP3008 10-bit eight-channel ADC. Just like our Arduino's internal ADC, this chip can read analog signals with a 10-bit resolution. It can be useful when your project needs more analog inputs than the Arduino has available. The MCP3008 needs to be connected to 5V and GND and its SPI interface needs to be connected to the corresponding SPI pins of the Arduino (pins 11, 12, and 13 for MOSI, MISO, and SCK, respectively). We use the Arduino Uno's pin 4 as the chip select pin, but any other pin not already in use would work just as well. The following schematic shows how to connect the MCP3008 to the Arduino Uno with the help of a breadboard and some jumper wires. In addition, a potentiometer is connected to the MCP3008's analog channel 0 input.

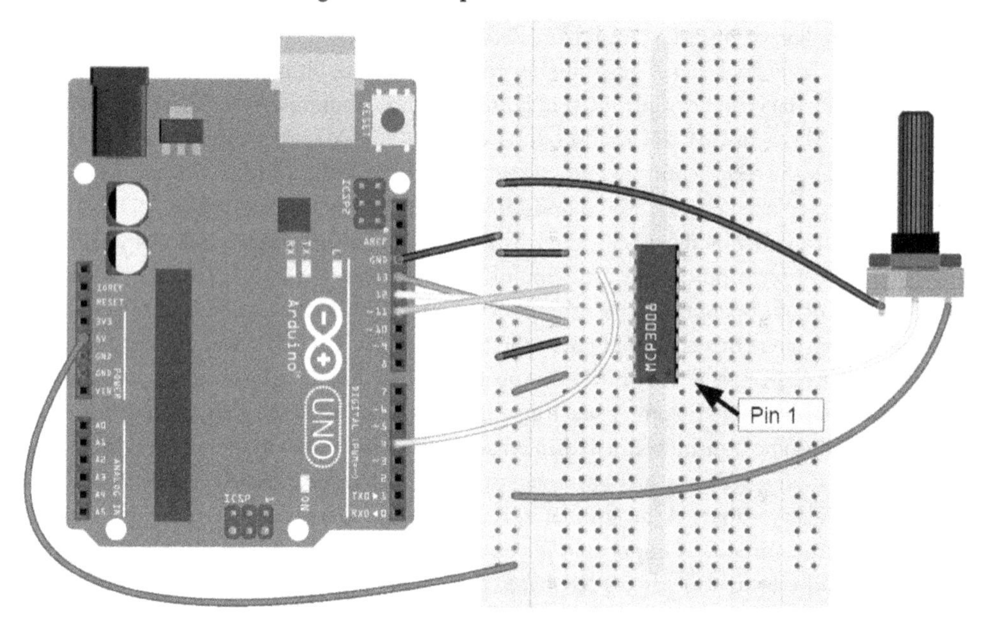


Figure 2.7 – Arduino connected to an MCP3008 ADC via SPI. A potentiometer is connected to channel 0 of the MCP3008

After including the SPI library and the usual initial setup of pin numbers and other constant parameters, we start the SPI interface with `SPI.begin()` in the `setup()` function and configure the chip select pin as output. Now, we can start to transmit data:

```
// Include SPI library.
#include <SPI.h>
// Using pin 4 for chip select signal.
const int cs_pin = 4;
// MCP setting for single-ended measurements.
const int mcp_set_sgl = 1;

void setup() {
  Serial.begin(115200);
  // Start the SPI bus.
  SPI.begin();
  // Configure chip select pin as OUTPUT.
  pinMode(cs_pin, OUTPUT);
}
```

The different steps of the SPI transaction are contained in the `mcp_read_channel()` function:

```
int mcp_read_channel(uint8_t channel) {
  if (channel > 7) {
    return -1;  // Return error code -1 if channel is invalid.
  }
  // Start the SPI transaction.
  // Bitrate is 1Mhz, bitorder is MSB first, SPI mode is 0.
  SPI.beginTransaction(SPISettings(1e6, MSBFIRST, SPI_MODE0));
  // Pull CS LOW to select the MCP3008.
  digitalWrite(cs_pin, LOW);
  // Transmit start bit.
  SPI.transfer(0x01);
  // Transfer settings bit and channel bits, padded with zeros.
  // Capture the returned byte in the msb variable.
  uint8_t msb = SPI.transfer(mcp_set_sgl << 7 | channel << 4);
  // Use a bitmask to extract the two payload bits.
  // These are the two most significant bits of the ADC value.
  msb = msb & 0b11;
```

```
    // Transfer a dummy byte and capture the returned byte.
    // This is the LSB of the 10 Bit ADC value.
    uint8_t lsb = SPI.transfer(0);
    // Push CS HIGH to release the MCP3008.
    digitalWrite(cs_pin, HIGH);
    // End the SPI transaction.
    SPI.endTransaction();
    // Combine the MSB and the LSB into the 16 bit return value.
    return msb << 8 | lsb;
}
```

It begins the SPI transaction with a set of settings that match the specifications of the MCP3008, as per its datasheet. It then pulls the chip select line low and the data transfer can begin. What data to transfer and how to interpret the results varies from chip to chip and is always described in the datasheet. In the MCP3008's case, we first transmit 1 byte that holds a start bit. We then transmit a byte that contains 1 bit of settings (1 for single-ended or 0 for differential input) and then the channel number we want to read (from 0 to 7). After two more clock cycles, the MCP30008 will transmit the analog value as 10 bits over the next 10 clock cycles. We capture these 10 bits in 2 bytes and then combine them into a single value that our function can return.

If you connect the output of a potentiometer to the MCP3008's channel 0 pin and open the Serial Plotter, you will see output that looks just like what we got from using `analogRead()`. Only this time, it is the MCP3008 and not the Arduino itself that does the analog-to-digital conversion for us, and the digitized data is transmitted to the Arduino via SPI.

Keep in mind that this code example is only meant to illustrate how an SPI transaction works under the hood. You will rarely have to write code that is as detailed and will usually find a library that is much easier to use.

### *UART Serial is a great debug interface*

The last serial interface we need to look at is the *Serial* interface of the Arduino, which uses the RS232 protocol to transmit data over the Arduino's built-in **Universal Asynchronous Receiver Transmitter** (**UART**) hardware. It is common to see the terms UART and Serial interface used synonymously, which makes the terminology a little fuzzy. Like the I2C bus, the Serial interface uses only two wires, but this is where the similarities already end. In contrast to I2C and SPI, UART is not a bus, has no controller or target, and has no dedicated clock line. The lack of a clock line (and thus a way to synchronize the receiver and transmitter) makes this an **asynchronous** interface. Instead of relying on a clock signal for synchronization, the transmitter and receiver must use a common, known data rate (and control their data rate very precisely), or otherwise, the transmission will be corrupted. For Serial interfaces, the data rate is called the **baud rate**. This is why we need to select the correct baud rate in the Arduino IDE's serial terminal, and selecting the wrong baud rate will cause us to either not see anything or see corrupted *garbage* data (feel free to try it out with the blink program from *Chapter 1, Introducing Robotics and the Arduino Ecosystem*). The two connections of a serial device are labeled **Transmit** (**TX**) and **Receive** (**RX**).

> **Note**
>
> When connecting two devices over their Serial interface, it is important to cross the wires such that the TX output of one device goes to the RX input of the other. Missing this step can lead to a lot of time spent debugging.

The Serial interface is relatively slow and not often used for low-level sensors. It is usually used to connect devices to a serial terminal and transmit data as human-readable ASCII symbols, which can be great for debugging. One of the few examples of sensor types that commonly support serial output in ASCII format is GPS receivers. They output their data as human-readable NMEA sentences. Since GPS has a relatively low update rate, to begin with, the low-performance interface does not pose a serious limitation. It is important to keep in mind that transmitting data as ASCII characters (human-readable text) generates a lot of overhead and slows down data transmission significantly. As an example, the value 255 can be transmitted in binary format as 8 bits, or a single byte. If we instead send the three ASCII characters, `'2'`, `'5'`, and `'5'`, we need to send 3 bytes for the same information content, and one more byte as a delimiter, such as a comma or a line break. In addition, we need to write software that then interprets these characters as a number. This can be cumbersome, but if we need to do it anyway, the Arduino IDE provides us with a set of very convenient helper functions, namely `Serial.parseInt()` and `Serial.parseFloat()`, for this task. We will use them in a later chapter.

The Serial interface is great whenever you want to make it easy to see what data is being transmitted. Because USB-to-Serial converters are easy to find and well supported by most computer systems, it is the best choice when we need our Arduino robot to communicate with a Serial terminal.

As we saw in *Chapter 1, Introducing Robotics and the Arduino Ecosystem*, the Serial interface of the Arduino Uno is used for programming the microcontroller via the IDE, and for communicating with the IDE's Serial Monitor and Serial Plotter. Using the same interface for sensors or other peripherals at the same time is usually not possible or can cause unexpected issues. This is another reason for avoiding the use of the Arduino Uno's Serial interface as a sensor interface. If we need a Serial interface for this purpose, the Arduino IDE makes it easy to use any two digital pins as a Serial interface via the **SoftwareSerial** library. While software Serial uses significant resources of the Arduino's microcontroller and might slow your program down, it can be extremely handy for prototyping. Larger Arduino boards such as the Arduino Mega (which we will get to know later) have multiple hardware Serial interfaces and make it easy to avoid having to use software Serial.

In the following example, we use a software Serial port to read data from a GPS receiver and echo the received data byte by byte to the hardware Serial port to view it in the Serial Monitor. The GPS receiver is powered by 5V and its TX and RX pins are connected to the Arduino Uno's software Serial RX (pin 10) and TX (pin 11), respectively (remember to cross over RX and TX between two UART interfaces). You can choose any other two digital pins as software Serial pins, except pins 0 and 1, which are used for the hardware Serial.

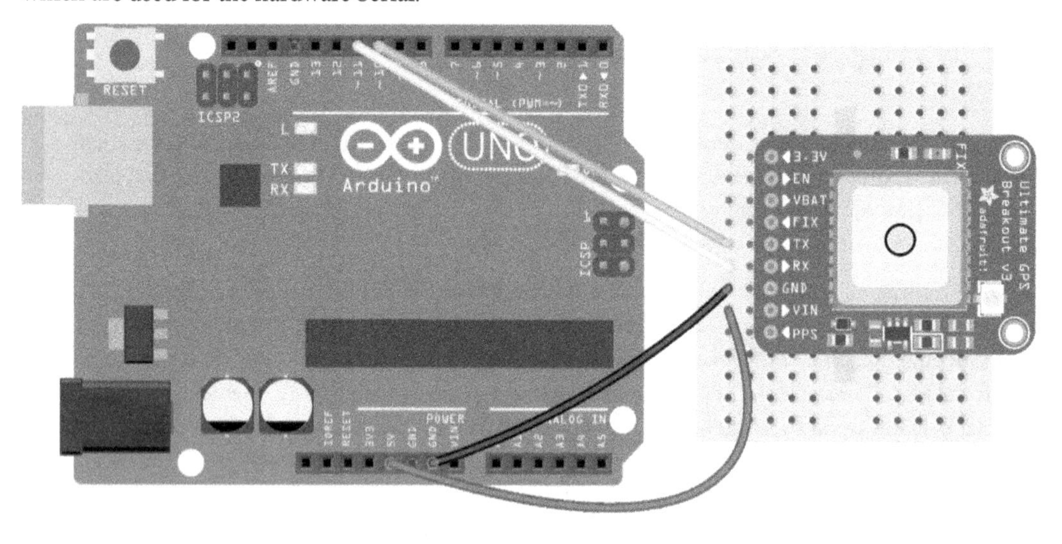


Figure 2.8 – Arduino Uno connected to a GPS receiver breakout board via a software serial port on pins 10 and 11

The GPS receiver used in the schematic shown in *Figure 2.8* sits on a handy breakout board from Adafruit (`https://www.adafruit.com/product/746`).

The following code instantiates a software Serial port on pins 10 and 11, periodically checks whether a byte was received from the GPS receiver, and if yes, echoes it to the hardware Serial port:

```
// Include the Software Serial library.
#include <SoftwareSerial.h>
// Create a Software Serial interface for the GPS receiver.
// It uses pin 10 for RX and pin 11 for TX.
SoftwareSerial gpsSerial(10, 11);

void setup() {
// Start the Hardware Serial interface.
  Serial.begin(115200);
  // Start the Software Serial interface.
  // GPS receivers use a baud rate of 9600.
  gpsSerial.begin(9600);
}
```

```
void loop() {
  // If we received a byte from the GPS receiver...
  if (gpsSerial.available()) {
    // ... print it to the Serial Monitor.
    Serial.write(gpsSerial.read());
  }
}
```

You may have noticed that we use `Serial.write()` instead of `Serial.print()` to echo the characters received from the GPS to the Serial terminal. `Serial.print()` converts numbers to their ASCII characters and then sends them. `Serial.write()` sends out the data as is. Since the GPS receiver already gives us all the characters in ASCII format, there is no need to convert them to ASCII with `Serial.print()`, so we can (in fact, we have to) use the simpler `Serial.write()` instead.

Figure 2.9 – The output of the GPS test sketch in the Serial monitor. Each line is a GPS NMEA sentence

You can tell from the output that there is some work left to do in order to parse useful information such as time, altitude, longitude, and latitude from the GPS sentences. You could either implement the parsing yourself as a fun exercise or use a readily available library such as Adafruit_GPS (`https://github.com/adafruit/Adafruit_GPS`).

## Additional peripherals for working with sensors

There are several types of devices that can make working with sensors easier in an Arduino project. Here is a brief list of a few that can come in very handy:

- **I2C multiplexers**: These chips take the I2C bus of a microcontroller and turn it into several, independent I2C buses. This can be incredibly useful when we want to use many sensors of a certain type, but the sensor only supports two different I2C addresses. Without an I2C multiplexer, we could only use two of these chips with our Arduino, but if we have a bus multiplexer (such as TI's TCA9548A) that splits our Arduino's single I2C bus into four buses, we can now use up to eight of the chips (two per bus) without risking address clashes. However, this will add additional overhead to the I2C communications, slowing them down further.
- **Differential I2C bus extender**: I2C is typically limited to a few centimeters of cable lengths. But what if you want to communicate with an I2C sensor that is located on the other side of a very large robot? A differential I2C bus extender (such as NXP's PCA9615) splits SCL and SDA into a pair of differential signals, greatly extending the admissible length of an I2C cable. Another extender converts the signals back from the differential to single-ended. With a pair of differential bus extenders and CAT5 cable in between, you can transmit I2C signals over up to 30 meters without losing performance.
- **I2C-to-SPI bridge**: If you want to use SPI devices but cannot afford to use the SPI pins of your Arduino, you can connect an I2C-to-SPI bus bridge (such as NXP's SC18IS602B) to the I2C bus and then communicate to the SPI devices over I2C. This of course eliminates all the performance advantages of SPI communication, and data transmission speed will be limited by the I2C bus.
- **External ADC**: If we need more analog inputs than the Arduino can provide, higher resolution, or higher data rate, we can use a dedicated external ADC chip, preferably with an SPI interface (such as the Microchip's MCP3008 from the SPI code example).
- **Logic-level shifters**: The Arduino Uno uses 5V logic, but many sensor devices only support 3.3V logic levels, and may even get damaged by a 5V signal. Fortunately, there are level shifter ICs (such as TI's TXB0108) that convert signals from 5V on one side to 3.3V on the other, and vice versa. Whenever you select a sensor, either make sure that it is compatible with the logic voltage level of your Arduino or consider using level shifters between the Arduino and the sensor.

There are many more helpful tricks that can support the use of sensors, for example, analog signal filters, buffer capacitors, and voltage dividers. But the peripherals here cover solutions to the most common problems that may arise when you start integrating various sensors into your robot.

## Summary

In this chapter, we got to know several useful types of sensors and the interfaces that an Arduino board can use to communicate with them. Needless to say that there is a lot more to write about each sensor type and the intricacies of each interface. But you now know the fundamentals and essential terms of the subject, which gives you a solid foundation from which you can dive deeper into any topic that is relevant to your next project.

After having learned a lot about the possibilities of adding sensing capabilities to your robot, the next chapter will be dedicated to motors that enable your robot to move.

## Further reading

You can find more information about the Arduino libraries and the components we used in this chapter in the following resources:

- The documentation of the Arduino Wire library (I2C): `https://www.arduino.cc/reference/en/language/functions/communication/wire/`
- The documentation of the Arduino SPI library: `https://www.arduino.cc/reference/en/language/functions/communication/spi/`
- The documentation of the Arduino Serial library (UART): `https://www.arduino.cc/reference/en/language/functions/communication/serial/`
- The documentation of Arduino methods to use external interrupt triggers: `https://www.arduino.cc/reference/en/language/functions/external-interrupts/attachinterrupt/`
- A detailed description of the Bosch BNO055 IMU (used in the I2C example): `https://www.bosch-sensortec.com/products/smart-sensors/bno055/`
- A detailed description of the Microchip MCP3008 ADC (used in the SPI example): `https://www.microchip.com/en-us/product/MCP3008`
- A deep-dive into SPI: `https://learn.sparkfun.com/tutorials/serial-peripheral-interface-spi`
- A deep-dive into I2C: `https://learn.sparkfun.com/tutorials/i2c`
- A deep-dive into asynchronous serial communication: `https://learn.sparkfun.com/tutorials/serial-communication`

# 3

# Making Your Robot Move and Interact with the World with Actuators

Every robot needs motors, or more generally, actuators, to move and interact with the world. This chapter will present a selection of robot actuators that are accessible, effective, and easy to integrate into Arduino-based robots.

We are going to look specifically at these types of actuators and learn how to use each of them in your Arduino project:

- RC servo motors
- DC motors
- Brushless DC motors
- Stepper motors
- Dedicated robot actuators

## Technical requirements

We will explore several different motor types and drivers along with the code to make them move with your Arduino Uno. You do not need the hardware to reproduce all of these examples, but at a minimum, you should have a standard RC servo on hand and follow along with the servo example. You will also need your trusty Arduino Uno again.

# Understanding electric motors and motor drivers

Even though there is a large variety of actuators you could choose from to power your robot (who would not want to build a rocket-powered robot at some point..?), by far the most common and versatile actuator type for DIY robots is the **electric motor**. This is why we will focus this chapter exclusively on electric motors. This leaves a lot to cover still since there are several different kinds of distinctly different electric motors that are useful for Arduino projects, all with their unique strengths and weaknesses. By the end of this chapter, you will understand what kinds of electric motor types you can choose from and when to use which.

Fundamentally, all electric motors that are relevant to DIY robotics work the same way. They contain electromagnets (copper coils or windings) that create a magnetic field when electric current flows through them and permanent magnets that are pushed around by this field. The motors are constructed such that this force between the two types of magnets is converted into torque around a motor shaft. The torque makes the shaft spin.

A microcontroller such as the ATmega328P on the Arduino Uno can drive an LED, but it can usually not drive an electric motor. The two main reasons are as follows:

- The motor windings require much higher currents to produce a meaningful torque than what a microcontroller alone can provide.
- The windings in motors have a large inductance (the electrical equivalent of mechanical inertia). This makes it hard to get current flowing, but even more importantly hard to rapidly stop current from flowing. We need special circuitry to avoid possible damage that driving inductive loads like motors can cause.

We, therefore, need a dedicated motor driver between the Arduino and the motor to handle the high current that flows from the battery of our robot into the motor. These motor drivers, at their core, have power transistors to switch the high currents and diodes to protect against damage caused by the inductance of the motor.

In order for a motor to spin, the electromagnetic field produced by the motor windings needs to constantly change. It needs to rotate around the shaft to keep pushing the magnets along their circular path. In order to achieve this, different windings need to be powered at different times during one motor revolution. This process is called **commutation**. The two main categories of electric motors that we are interested in, brushed DC and **brushless DC** (**BLDC**), differ in how this switching between powering different windings is realized. Because Arduino robots are generally battery-powered, we will only look into **direct current** (**DC**) motors that can be powered by batteries.

## Brushed DC motors

In brushed DC motors (usually referred to as simply DC motors), there are sliding contacts between the outer part of the motor, called the stator, and the motor shaft that has the windings attached to it called the rotor. This contact is made between brushes in the stator that slide over a segmented commutator on the rotor. The brushes are directly connected to the two motor terminals that can be powered directly from a battery to make the motor spin. The motor windings on the rotor are powered by the commutator and as the motor rotates different segments of the commutator touch the brushes of the stator, and different sets of corresponding windings are energized. This **mechanical commutation** happens at exactly the right time to create the rotating magnetic field that makes the motor spin.

Figure 3.1 – The rotor of a brushed DC motor

In *Figure 3.1*, you can see how the copper windings in the center are connected to the commutator segments on the right over which the brushes of the stator slide. The stator, motor housing, and brushes are not shown in this picture.

### Brushless DC motors

As the name brushless suggests, BLDC motors do away with the need for brushes. BLDC motors have three wires coming out of them so they cannot be directly connected to a bipolar DC power supply or a battery. Instead, BLDC motors are always paired with a complex driver circuit called an **electronic speed controller** (**ESC**). The ESC takes power from a battery and controls the current flowing through the three wires going into the motor such that a rotating magnetic field is created and the motor spins, a process called **electronic commutation**.

There are more types of brushless motors than just high-power BLDC motors. The most important ones for robotics that we will also take a closer look at later in this chapter are stepper motors. These have four wires coming out of them and are much easier to drive than three-phase BLDC motors.

## Different motor types and how to use them

Now that we have a good understanding of the basics of how electric motors for DIY robots work, let us look in more detail at the most used motor types. We will discuss the advantages and disadvantages of each of these motor types and go through the hardware setup and Arduino code to use them.

## Using RC servo motors

The easiest electric actuator type to use with Arduino is the standard RC servo motor. An RC servo is an integrated system that has the following components, all packaged in a very compact housing:

- An electric motor with a gearbox, most commonly a miniature DC motor
- A suitable motor driver circuit
- A shaft encoder at the output of the gearbox, most commonly a simple potentiometer
- A microcontroller that reads an input signal and controls the position of the gearbox output by driving the motor accordingly

There is a large selection of RC servos available that vary in size, speed, torque, and input voltage. Most commonly, RC servos run on 5V, have a range of motion of around half a turn, and move slowly but with large torque thanks to their high gear reduction. The most common interface for RC servos is a **pulse width modulated** (**PWM**) input signal over a single wire. To control the servo, this pin needs to be periodically pulled high and low, at a rate of 50 Hz. The desired output position of the servo is encoded in the duration of the high pulse (the pulse width), with 1 ms corresponding to the minimum position and 2 ms to the maximum position. These values vary somewhat between servo types. An RC servo has three wires: one for 5V power (usually red), one for GND (usually black or brown), and one for the PWM input signal (usually orange, yellow, or white). We can connect a servo motor directly to the Arduino by connecting its 5V and GND wires to the corresponding pins on the Arduino, and the signal wire to any pin that has a tilde (~) symbol in front of the pin number. On the Arduino Uno, these are pins 3, 5, 6, 9, 10, and 11. These pins can generate a PWM output without using any CPU capacity, thanks to dedicated timer hardware. The Arduino documentation calls them *analog output pins*, which is a little misleading.

Servos are great whenever we do not need continuous rotation or particularly fast motion. You can use them to power the joints of a robot arm or the legs of a legged robot, move the pen of a pen plotter up and down, operate valves for watering plants, or automatically lock a door's deadbolt. Since the PWM signal that controls them is updated only at 50 Hz, they are not suited as actuators in control loops that require a faster control rate than that.

The Arduino IDE provides us with the `Servo.h` library, which makes controlling a servo extremely easy. At its core, the library provides the `writeMicroseconds()` function, which sets the duration of the high pulse. After attaching a servo to one of the PWM-capable pins, we can use this function to control its position. In the following example, we use the Arduino to control a servo motor according to the position of a potentiometer. The hardware setup is shown in *Figure 3.2*:

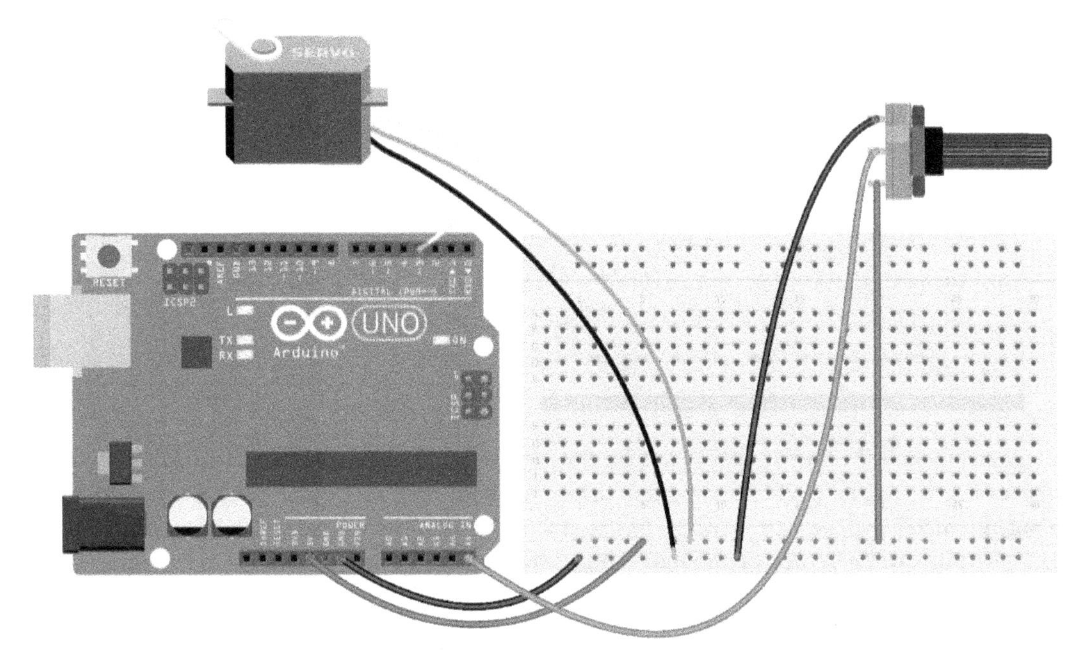


Figure 3.2 – Arduino Uno with an analog potentiometer as input on pin A5 and an RC servo controlled by the PWM signal of pin 3

The code for this example is as follows. Try to understand every line on your own before you read the explanation that follows:

```
// Include the Servo library.
#include <Servo.h>

// Analog input pin.
const int input_pin = A5;
// Servo output pin. Must be a PWM pin.
const int servo_pin = 3;
// Instantiate a Servo object called servo.
Servo servo;

void setup() {
  // Attach the servo object to the servo pin.
  servo.attach(servo_pin);
}
```

```
void loop() {
  // Read the potentiometer.
  int input = analogRead(input_pin);
  // Map the analog value from 0 to 1023 to a pulse width
  // between 1000 and 2000.
  // The suffix _us stands for microseconds.
  int pulse_width_us = map(input, 0, 1023, 1000, 2000);
  // Set the servo PWM pulse width to that value.
  servo.writeMicroseconds(pulse_width_us);
}
```

We will first include the `Servo` library and instantiate a servo object. In the `setup()` function, all we have to do is attach a PWM output pin to the servo object. In the `loop()` function, we sample the potentiometer, use the `map()` utility function to map the 10-bit analog input value (from `0` to `1023`) to a pulse width between a minimum and a maximum, and command the servo by setting the pulse width with `writeMicroseconds()`.

If your Arduino runs this sketch and everything is wired up correctly, you can turn the potentiometer, and your servo will move accordingly. This will work fine with small servos that can move freely, even if your Arduino is powered only via USB. If you use a bigger servo or put a mechanical load on it, the servo might draw more current than the USB port can supply. In this case, you will need to power your Arduino from a more capable power source.

Any RC servo should be able to handle a pulse width between 1,000 and 2,000 microseconds. To get the maximum range of motion out of your servo you can likely lower the minimum pulse width and raise the maximum pulse width by a few hundred microseconds. However, at some point, your servo will hit its internal hard stops, which can potentially damage it.

If you want to use more servos than your Arduino has PWM output pins, you can use dedicated PWM generator chips such as the NXP's *PCA9685*. This useful chip offers 16 independent PWM output channels and can be controlled by your Arduino via I2C.

## Using brushed DC motors

DC motors are very easy to use, as all they need is a DC voltage source such as a battery. The motor handles the required switching between the windings to create a rotation by mechanical commutation. A driver circuit to control the speed of a DC motor can be built very easily from just a few components, namely transistors, resistors, and diodes. There are even a lot of ready-to-use, integrated motor drivers available. Because they are so easy to use and available in an incredible variety, you can find DC motors in most applications where electric motors are required, from toys and power tools to appliances and even some electric vehicles. For the same reasons, they are also an excellent choice for your DIY robots. However, there are a few downsides to this motor type that are a direct consequence of the need for brushes. The brushes wear out over time, especially when the motor operates at high speeds or high currents. This effectively limits the amount of current we can push through a DC motor, and thus its power output. The brushes and commutator also take up space and add weight, making the motor bigger and heavier. All these factors decrease the maximum **power density** of a DC motor. In addition, the fast mechanical switching that happens at the brushes when the motor spins in combination with the high winding inductance produces constant sparking at the commutator. This **brush sparking** can be a significant source of EMI, distorting analog signals of nearby sensors or interfering with the wireless interfaces of your robot. The heat generated by the friction of the sliding brushes along with the brush sparking reduces the energy that the motor can convert into mechanical energy, limiting its efficiency.

Whenever an RC servo does not meet your requirements, trying to find a suitable DC motor is often the next step. It is quite common to pair DC motors with reduction gearboxes to reduce the speed of the motor and increase the torque. There are many integrated geared motors (a combination of motor and gearbox) available. Even though a gearbox reduces efficiency even further, using a geared DC motor is often the easiest way to create the drive train for a moving robot. If your application requires higher power density or more precise open-loop control, you will need to consider using brushless or stepper motors.

The most common way to control a DC motor is with a special circuit called an **H-bridge** or full-bridge. It is easy enough to build an H-bridge from discrete components on a breadboard, but using an integrated circuit that implements an H-bridge in just one component, such as TI's *L293D* or ST's more capable *L298*, is even easier.

Now that we know the advantages and disadvantages of brushed DC motors, let us look at how to use them with Arduino. In this example, we will use the *Arduino Motor Shield Rev3* together with our Arduino Uno to drive a 6V DC motor. We will again use a potentiometer as an input device to control the direction and speed of the motor. The hardware setup is shown in *Figure 3.3*:

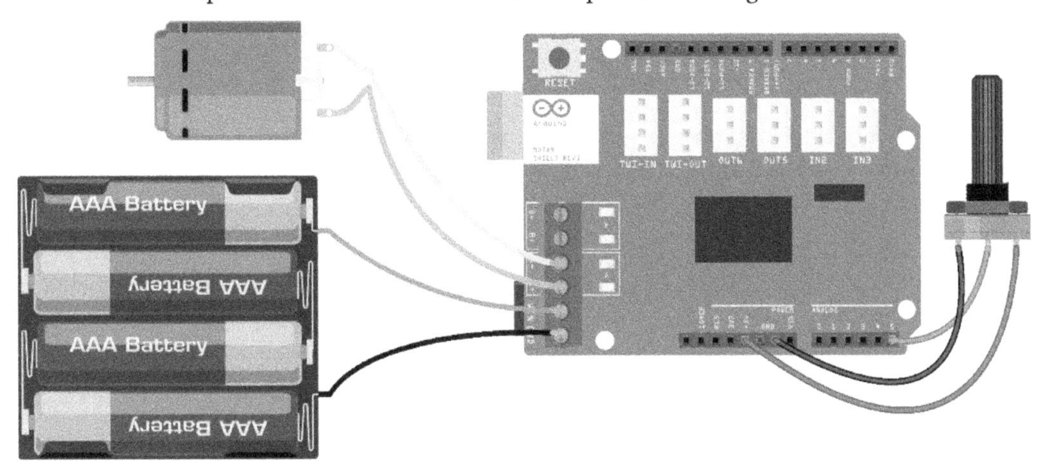


Figure 3.3 – Arduino Uno with motor shield connected to a battery, a DC motor, and a potentiometer as the input device

The motor shield is built around the *ST L298* dual full-bridge driver. This chip can independently drive two DC motors. Like every Arduino shield, it fits right on top of the Arduino Uno and exposes all its pins on the top. However, some of the pins are now used by the motor shield and should not be used for anything else. These pins are marked on the top of the shield with their specific function (for example, *DIRA* and *PWMA*).

> **Note**
>
> Be careful when inserting the shield into the Arduino to not bend any of its long pins!

The external battery is connected to the **Vin** and **GND** inputs. The motor is connected to **A+** and **A-** outputs. We use a potentiometer as an input element, connected to analog pin A5 on the motor shield. If you use a small motor that does not need a lot of power to spin, you do not even need an external battery. In this case, the motor will be powered from the USB port of your computer. For most motors big enough to move a small robot, USB power will not be enough though.

To control the motor via the motor shield, the Arduino needs to command the direction via the DIR pin. If the pin is high, the motor will spin in one direction (let us call this the forward direction), and if it is low, it will spin backward. The Arduino also needs to command the motor speed via the PWM pin. We can use the `analogWrite()` function with arguments from `0` to `255` to set PWM signals from always low to always high, with every duty cycle in between. The L298 driver chip on the motor shield controls which side of the battery gets connected to which motor lead according to the status of the DIR pin. It switches the motor on when the PWM pin is high and off when it is low. The motor transforms this rapid switching into smooth rotation at a certain speed (roughly proportional to the duty cycle), thanks to its large inductance and mechanical inertia. The motor shield also supports a brake function. However, this is rarely useful in practice, and we therefore simply disengage the brake in the `setup()` function of our program by pulling the brake pin low.

> **Note**
>
> *Brake* does not refer to a mechanical brake here. Instead, the motor shield can slow down the motor by short-circuiting its two terminals. This will make the motor much harder to turn thanks to an effect called electrodynamic braking.

The code for this example reads the potentiometer and maps one side of its range (`0` to `512`) to motor speeds in one direction, and the other side of its range (`512` to `1023`) to speeds in the other direction by manipulating the value of the direction pin accordingly. This happens in an `if/else` statement inside `loop()`. This statement might look a little confusing, so *Figure 3.4* illustrates the logic:

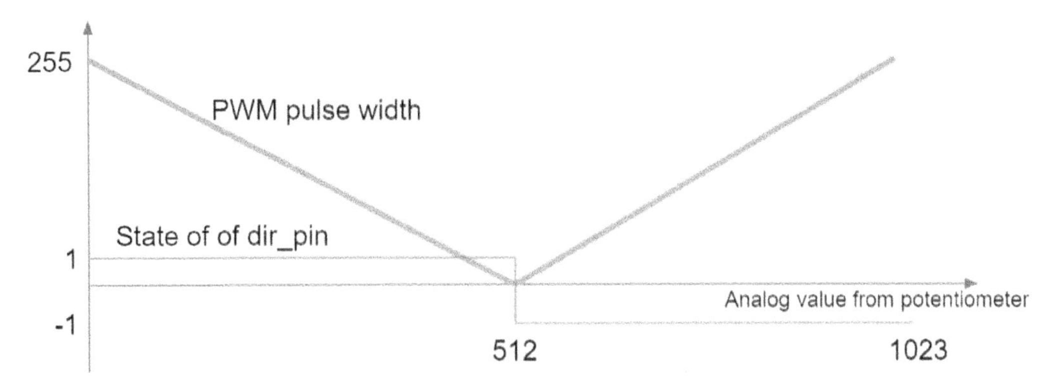


Figure 3.4 – Mapping from an analog input pin to a direction pin and PWM value for bidirectional motor control

The motor shield can control two motors, one on channel A and another one on channel B. In this example, we are only using channel A:

```
// Define the pin numbers.
const int dir_pin = 12;
const int brake_pin = 9;
const int pwm_pin = 3;
const int analog_pin = A5;

void setup() {
  // Set direction and brake pins as outputs.
  pinMode(dir_pin, OUTPUT);
  pinMode(brake_pin, OUTPUT);
  // Disable brake functionality.
  digitalWrite(brake_pin, LOW);
}

void loop() {
  // Read the potentiometer voltage.
  int analog_val = analogRead(analog_pin);
  // Map the potentiometer value to
  // direction and PWM output (see Figure 3.4).
  if (analog_val < 512) {
    // Turn forward.
    digitalWrite(dir_pin, HIGH);
```

```
      int pwm;
      pwm = map(analog_val, 0, 511, 255, 0);
      analogWrite(pwm_pin, pwm);
    } else {
      // Turn backward.
      digitalWrite(dir_pin, LOW);
      int pwm;
      pwm = map(analog_val, 512, 1023, 0, 255);
      analogWrite(pwm_pin, pwm);
    }
}
```

The standard PWM frequency on the Arduino PWM pins is 490 Hz, which is fairly low for DC motor control and might lead to audible noise generated by the motor, or even to motor vibrations. There are ways to increase the PWM frequency to mitigate these issues, but they have side effects that go beyond what we can discuss in this chapter. We will revisit this issue and learn a solution for this problem in *Chapter 13, Building a Self-Balancing, Radio-Controlled Telepresence Robot.*

## Using BLDC motors

Thanks to the lack of brushes, brushless motors have several advantages over DC motors. They are virtually free of wear and tear, we can push very large currents through them to achieve high power outputs, they can be built small and power-dense, and they have exceptional efficiency. The main downside of being brushless is that the continuous switching between windings to create a rotating magnetic field (commutation) needs to happen some other way. The windings of BLDC motors are connected internally so that they have three wires coming out of them. This means they cannot be directly connected to a DC power supply such as a battery to make them spin. Instead, BLDC motors are paired with a relatively complex driver circuit, for example, an ESC.

## ESCs for BLDC motors

The ESC takes power from a battery and controls the current flowing through the three wires going into the motor to create the rotating magnetic field that makes the motor spin. Orchestrating the current into the three motor wires to generate the correct magnetic field at every point in time to achieve maximum efficiency is not an easy task. To achieve this, the ESC has a microcontroller itself that runs special software to estimate the state of the motor and control the ESC's power transistors accordingly. Building an ESC to control a BLDC motor is much more complicated than building a driver for a brushed DC motor. Unless you want to learn how to build one yourself, it usually makes sense to use an off-the-shelf ESC.

The need for an ESC makes brushless motors more expensive and more complicated to use than brushed motors. But whenever we need high power density, high efficiency, longevity, or low EMI emissions, they are the best choice for your robot. The availability of cheap BLDC motors with efficient ESCs has enabled many new kinds of DIY robots, with high-performance quadcopter drones being the most prominent ones. Certain types of brushless motors (often used as gimbal motors) are built for low-speed, high-torque operations. When they are paired with the right ESC, they can often be used instead of a mechanically more complex geared DC motor.

Because brushless motor controllers are complex devices in and of themselves, with microcontrollers that enable a lot of different functionalities, there are several distinct types of them. At a high level, they can be classified as either open-loop **speed controllers** or complete, closed-loop **servo controllers**. Basic speed controllers are sensorless, and higher-end ones support sensored operation.

Let us look at all these types of BLDC controllers in a little more detail.

### *Sensorless ESCs*

An ESC needs to know the current rotor position to drive the motor correctly. A sensorless ESC, the most basic and most common BLDC controller for hobby and DIY applications, needs to determine the motor position without sensors. This makes it cheap and simple but brings with it several limitations. If the motor is spinning, the ESC can use the motor itself as a rudimentary position sensor (keyword *back EMF*). But if the motor is standing still, the ESC has no chance of determining the rotor position. This means that a sensorless ESC cannot smoothly start a motor from a standstill without some amount of initial jerk or jittering, and it does not allow for smooth or efficient motor control at very low speeds. At higher motor speeds, however, sensorless ESCs can perform great. This limitation is not a problem at all for drones or airplanes where the motor is used at high speeds for the entire time of flight. But it is a problem for wheeled mobile robots that stop and start frequently and that need to accurately control their wheels at low speeds.

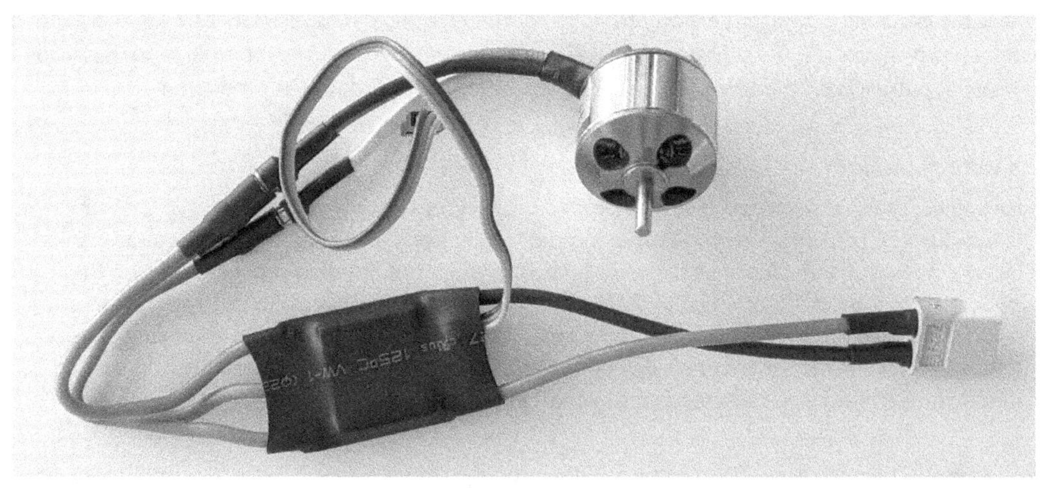

Figure 3.5 – A small, sensorless BLDC motor with a matching ESC

#### Sensored ESCs

To enable smooth startup and operation at low speeds, higher-end BLDC motors are equipped with three **Hall sensors**. These sensors are used as a special form of encoder that tells the ESC just enough information about the rotor position to let it drive the motor correctly, independent of the motor speed. Sensored ESCs can read the Hall sensor signals and deliver great motor control from a standstill and at low speeds. This makes the combination of BLDC motors with Hall sensors and sensored ESCs a great (but expensive) choice for driving the wheels of a powerful mobile robot.

#### Servo controllers

One step up from sensored ESCs are BLDC servo controllers. On top of driving the motor windings and reading Hall sensor signals, a servo controller also has interfaces to shaft encoders. A BLDC motor with Hall sensors and a shaft encoder, paired with a servo controller, creates a powerful servo motor that you can command to move to a certain shaft position at high precision or rotate at a certain speed. A BLDC servo motor system is the most capable (but also most complex and expensive) actuator concept that you can consider for your robot if you need the power density and compactness of a BLDC motor, paired with very precise controllability.

### Driving a BLDC motor with a sensorless ESC

After all this theory, let us now drive a simple sensorless BLDC motor with our Arduino Uno using a sensorless ESC. You can buy a set containing a small BLDC motor and matching ESC at around the same cost as the Arduino motor shield. Hobby ESCs are made for use in RC applications, so they have the same 50 Hz PWM input that we know already from working with RC servo motors. This means we can simply reuse our servo sketch from earlier, but instead of a servo, we can connect the three signal wires of the ESC (5V, GND, and PWM input) to our Arduino.

Because the exact minimum and maximum pulse width of the RC PWM interface can vary between different receivers, most ESCs support a *calibration* step that allows you to calibrate its input range to the range of pulse widths that the RC receiver, or in our case, the Arduino, is producing.

> **A word of caution**
>
> During the calibration process, you will need to connect the battery (for example, a 3S Lithium battery) to the ESC, making the setup potentially dangerous. Even small BLDC motors can be very powerful and can cause harm and damage when not handled properly. During the calibration process, unexpected rapid motor movement can occur. Make sure that your motor is securely attached to a stand such that it cannot easily break free and that none of its moving parts touch wires, your skin, or your hair. Do not hold it in your hand. This is especially important for outrunner motors, such as the one shown in *Figure 3.5*.

With the RC servo sketch from the beginning of this chapter loaded onto your Arduino, the ESC connected to the Arduino, and the BLDC motor connected to the ESC, we are ready for calibration!

If you have a manual for your ESC, it likely explains the PWM calibration procedure. If your ESC did not come with a manual, a common calibration procedure across many ESCs is as follows:

- Make sure the battery is disconnected from the ESC.
- Turn the potentiometer that is connected to the Arduino all the way to 5V such that your Arduino outputs the PWM signal with the longest pulse (indicating maximum speed to the ESC).
- Turn on the ESC by connecting the battery to it – be careful, the motor might spin unexpectedly anytime now (even though it is not supposed to).
- The motor should not spin at this point. Wait until you hear a beep signal from the motor.
- Turn the potentiometer all the way in the other direction (indicating motor off to the ESC).
- Wait until you hear another beep signal from the motor.

At this point, the ESC is calibrated to the pulse width range of the input signal and can be used. If you carefully turn the potentiometer, the motor should start spinning and get faster the farther you turn the potentiometer. You will notice the jerky motor start, which is due to the fact that this setup does not have Hall sensors.

There is typically no need to perform the calibration procedure each time the ESC is turned on. It saves the PWM range parameters to permanent memory. To avoid accidentally entering the calibration procedure, make sure the PWM signal commanded to the ESC when connecting the battery is the motor-off signal (shortest pulse width). If you turn on the ESC with the potentiometer somewhere in between the motor off and full speed setting, the ESC should not spin the motor. Only when you turn the potentiometer all the way to the off position will it arm the motor and give a beep signal. From there on, you can control the motor as usual by turning the potentiometer. There is of course no need for a potentiometer to make the motor spin. Any program that you write for your Arduino can drive the motor by simply commanding a PWM signal to ESC.

Simple sensorless ESCs cannot change the direction of rotation. If you want the motor to spin the other way, you need to swap any two of the three connections between the ESC and the motor.

The majority of small ESCs have a **battery elimination circuit** (**BEC**) built into them. This means they provide 5V to the Arduino (or RC receiver) over the red cable in the center of the three-pin connector. With the battery connected to the ESC, you can unplug your Arduino from the USB port, and if your ESC has a BEC it will seamlessly continue to work, now powered from the ESC's battery.

## Using stepper motors

Stepper motors are another kind of brushless motor. They are similar to BLDC motors but designed for quite different use cases. While BLDC motors are optimized for power density, stepper motors are optimized for *large holding torque* and *accurate open-loop controllability*. This means they are ideal for applications that need very precise control of the motor position without the need for a position sensor or a feedback controller. Instead, if we want a stepper motor to rotate a certain number of steps

(200 steps per rotation is a common resolution), we simply command that many steps to the motor and can trust that it precisely turns accordingly. If you want it to rotate by 45 degrees (an eighth of a full rotation), for example, all you need to do is command 25 steps. Stepper motors are commonly found wherever precise motion is required, but neither speed, power, nor low weight is a priority. You can often find stepper motors paired with belt drives or ball screw transmissions in printers where they move the paper and the printhead, in flatbed scanners where they move the scanner head, and in 3D printers and CNC machines. Due to their power limitations, they are not suitable for powering drones, airplanes, or fast mobile robots.

Even though stepper motors are brushless motors, their internal wiring is different from BLDC motors. The most common wiring type for steppers creates a **bipolar stepper motor** that has four wires coming out of it (instead of three like in BLDC motors). Two of the four wires feed one set of internal windings and the other two feed another independent set of internal windings. We will only look at how to work with bipolar stepper motors in this chapter and skip over the **unipolar stepper motors** that have more than four wires. If you know the pinout of a unipolar stepper, you can connect some of its wires to turn it into a bipolar motor, and then use the control method that this section will introduce. Stepper motors are often made with standardized mounting interfaces, called *NEMA* XX. *Figure 3.6* shows a bipolar stepper motor with a NEMA 17 mechanical interface.

Figure 3.6 – A NEMA 17 bipolar stepper motor from the extruder of a 3D printer

A bipolar stepper motor has two sets of internal windings, called *phase A* and *phase B*. We can drive current through each of these two phases in two directions, *positive* (+) and *negative* (-). This means there are four *states* we can operate the motor in:

| State | Phase A current direction | Phase B current direction |
|---|---|---|
| 0 | + | + |
| 1 | - | + |
| 2 | - | - |
| 3 | + | - |

Table 3.1 – The four operating states of a bipolar stepper motor

To make the motor move one step, we need to change the state, that is, the current direction in one of the phases, according to this table. If we go one line down in this state table, the motor will go one step forward; if we go up one line, it will step backward. To make the motor spin, we just need to step through these states in quick succession to create a smooth motion. The faster we step, the faster the motor spins. This is quite different from driving a DC motor, where we made use of a PWM signal to turn the motor on and off quickly to control the speed. Turning a stepper motor on and off rapidly will not affect its speed (since speed is dictated by the frequency of steps), but it will still control the current that flows through the motor and, thereby, its (holding) torque. Consequently, there are two things we can control separately: speed and torque. In practice, we will try to feed just enough current to the motor (by having a high enough PWM duty cycle) so that it successfully makes every step we command. If we use more current, we drain our robot's battery faster and, often more importantly, risk overheating the motor or driver. If we use less current, the motor is not able to perform every step we command, a phenomenon called **losing steps**. Since we blindly trust the motor to make every step, for example, to produce a high-quality 3D print, we cannot compensate for lost steps. In practice, a stepper motor that is losing steps can be a troublesome problem to debug.

To drive a bipolar stepper motor with our Arduino Uno, we can use the same motor shield that we used for the DC motor example. It conveniently has two output channels that we can use for the two sets of windings in the stepper motor, and it lets us control the direction and current (via PWM) of each channel individually. The setup looks remarkably similar to the DC motor experiment, as you can see in *Figure 3.7*. We add a second potentiometer as an input so that we can control both speed and current, and we simply connect the stepper motor as if it were two DC motors. If you have used simple AA batteries so far, it is time to switch to a rechargeable battery between 6V and 12V for this example. Even small stepper motors tend to draw more current than alkaline AA batteries can deliver.

If you do not know which two pins of your motor belong to the same phase, there are several ways to find out. The easiest is to create a short circuit between any two of the wires. If that makes the motor noticeably harder to turn by hand, you have found a matching pair. This is electrodynamic braking in action!

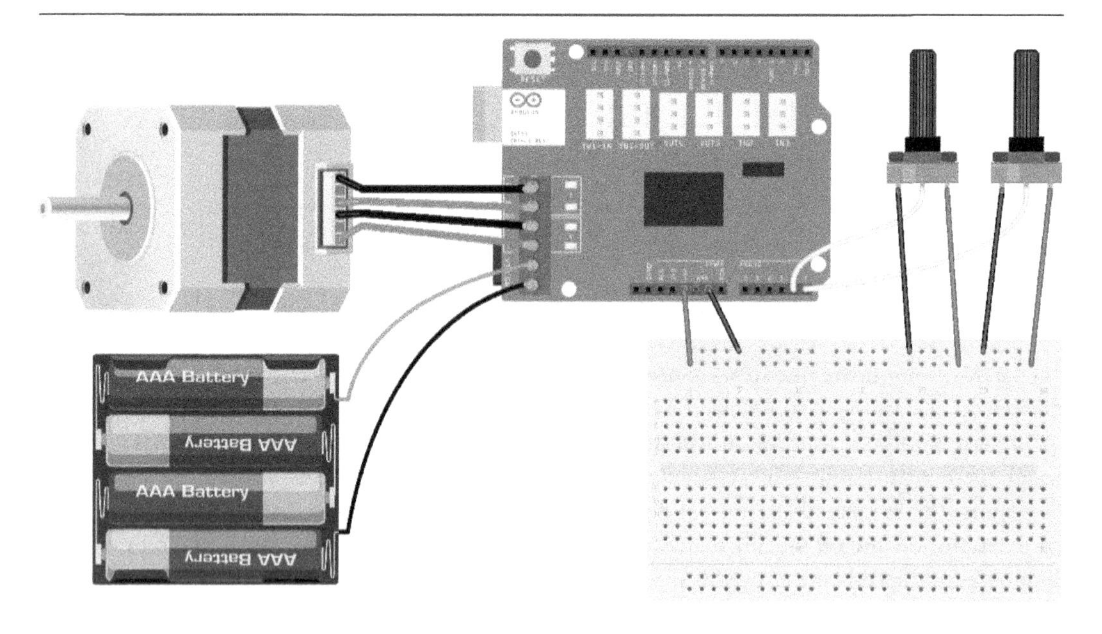


Figure 3.7 – An Arduino Uno with the Arduino motor shield, a bipolar stepper motor, a battery pack, and two potentiometers as control inputs

The Arduino IDE provides us with the `Stepper.h` library that makes using a stepper motor with the motor shield quite easy. Using this library, we can easily write code to control the speed and torque of this motor.

As usual, we first include the library and declare the `const` parameters to hold the relevant pin numbers. We can also declare a `const` parameter that holds the number of steps per revolution of our motor. For most stepper motors, this is 200:

```
// Include stepper motor library.
#include <Stepper.h>
// Define the pin numbers for phase A.
const int dir_pin_a = 12;
const int brake_pin_a = 9;
const int pwm_pin_a = 3;
// Define the pin numbers for phase B.
const int dir_pin_b = 13;
const int brake_pin_b = 8;
const int pwm_pin_b = 11;
// Analog input pins.
const int analog_pin_power = A5;
```

```
const int analog_pin_speed = A4;
// Most steppers have 200 steps per revolution.
const int stepsPerRev = 200;
```

We then instantiate a stepper object of the Stepper library, passing in the motor parameter and the two pins that control the direction of the current in the two sets of windings:

```
Stepper bipolarStepper(stepsPerRev, dir_pin_a, dir_pin_b);
```

In the `setup()` function, we set the control pins as outputs and disable the brake functionality for both channels:

```
void setup() {
  // Set control pins as outputs.
  pinMode(dir_pin_a, OUTPUT);
  pinMode(brake_pin_a, OUTPUT);
  pinMode(dir_pin_b, OUTPUT);
  pinMode(brake_pin_a, OUTPUT);
  // Disable brake functionality.
  digitalWrite(brake_pin_a, LOW);
  digitalWrite(brake_pin_b, LOW);
}
```

In the `loop()` function, we sample both potentiometers and map their analog values to a motor speed (in rotations per minute) and a PWM duty cycle between `0` and `255`. We will then use the `setSpeed()` library function to set the speed and, if it is greater than `0`, use the `step()` library function to command the stepper motor to take a step:

```
void loop() {
  // Map power potentiometer to a PWM output.
  int power_input = analogRead(analog_pin_power);
  int output_pwm = map(power_input, 0, 1023, 0, 255);
  // Map speed potentiometer to an rpm output.
  int speed_input = analogRead(analog_pin_speed);
  int rpm = map(speed_input, 0, 1023, 0, 100);
  // Drive both phases with the same PWM.
  analogWrite(pwm_pin_a, output_pwm);
  analogWrite(pwm_pin_b, output_pwm);
  // Set the motor speed.
  bipolarStepper.setSpeed(rpm);
```

```
    // Only call step() if rpm is greater than 0.
    if (rpm > 0) {
      bipolarStepper.step(1);
    }
  }
```

The `step()` function will take as much time as it takes to do the step. This behavior is in general very undesirable as it can **block** for a long time (a step at zero speed would block the program forever). If we were using this stepper motor in a robot, we would want to use a third-party library that implements a non-blocking `step()` function or write our own. But for this small example where performance does not matter, the Arduino Stepper library has the benefit of being extremely easy to use.

### Experimenting with the stepper motor

If the example code runs on the hardware shown in this example, you can control the torque and the speed of the motor individually. If you try to stop the motor by hand, you will notice that it is much easier to stop when you turn down the power via the power potentiometer. You will also notice that, regardless of the power setting, the motor becomes easier to stop at higher speeds. This is something to keep in mind when building robots with stepper motors. The faster they move, the more likely they are to lose steps. Because stepper motors always draw significant currents, even at low speeds and at standstill, you might notice that the L298 driver chip gets hot during operation.

> **Caution**
>
> The stepper motor and the driver can both get very hot. Use caution when touching them.

### Other ways of driving a stepper motor

Using the Arduino motor shield and the Arduino Stepper library as we did in this section is the most basic way of driving a stepper motor. What makes it so basic is that it uses *full steps* to drive the motor. If you modify the code to command exactly `stepsPerRevolution (200)` steps, your motor will turn exactly one rotation. Using more sophisticated methods, it is possible to drive stepper motors in smaller steps, called **microsteps**. If we drove our motor with 8 microsteps per step, for example, we could achieve 1,600 distinct microsteps per revolution, and thus much higher position resolution and smoother rotation. There is a wide selection of stepper driver chips and boards available that can do this for us. You can often set the microstep resolution and the current limit in hardware (via solder bridges and a potentiometer), and then use the driver chip to control the motor accordingly. The most common interface for these stepper driver ICs is a **step/direction input**. The direction input (high or low) determines the direction, and a high/low pulse to the step input pin makes the motor turn one (micro) step. Some of these driver chips have additional features such as incredibly quiet motor operation or adaptive current settings to keep the motor and driver as cool as possible, while still avoiding losing steps.

Figure 3.8 – A small, integrated stepper motor driver board

In the photo of the stepper driver board in *Figure 3.8*, you can see the heat sink and the direction input pin of the step/direction interface labeled **DIR** in the bottom-right corner. You can also see the potentiometer for setting the motor current in the top left corner. Stepper driver boards with this form factor are commonly found in 3D printers.

# Dedicated robot actuators and servo controllers

The RC servo we got to know earlier in this chapter is a self-contained actuator but it has significant limitations. Most importantly, it does not provide any feedback (for example, position, speed, torque, and internal temperature), it does not rotate much more than half a rotation, and it only has one mode of operation, namely position control.

There are several dedicated robot actuators that integrate a motor, gearbox, motor controller, position, torque, and temperature sensors, as well as powerful interfaces for rich feedback. These actuators often offer different modes of operation that allow you to either a target position, a target speed, or a target torque. Most of these integrated actuators tend to be too expensive or too complicated to be widely used in DIY Arduino projects. The most relevant robot actuators for DIY robots nowadays are the *Dynamixel* servos by *Robotis*. These devices are capable servo motors for high-end Arduino robots but they are an order of magnitude more expensive than the motors we have discussed so far.

Alternatively, you can purchase a servo controller that you can combine with the motor and position sensors of your own choice to form a custom servo motor. This can be an interesting avenue for high-performance robots but it also tends to be expensive and complex. A comparatively low-cost yet high-performance servo controller for BLDC motors is the *ODrive* by *ODrive Robotics*.

You can also relatively easily build your own high-performance servo motor using an Arduino Uno, an Arduino motor shield, and a geared DC motor with an encoder. We will do this as an example of a feedback control system in *Chapter 5, Getting Started with Robot Programming*.

## Summary

In this chapter, you learned how to use RC servos, DC motors, BLDC motors, and stepper motors with your Arduino and the Arduino motor shield. We wrote example code using the *Servo* and *Stepper* libraries that are built into the Arduino IDE. You now have a good understanding of the strengths and weaknesses of different actuator choices and can choose the right one for every aspect of your next robotics project.

After learning a lot about using the Arduino Uno for the core robot functionalities (sensing and acting), the next chapter will focus on the key features of other commonly used Arduino boards and why they can be the right choice for your robot.

## Further reading

The following resources can help you dive even deeper into the topics we discussed in this chapter:

- Arduino Servo library reference: `https://www.arduino.cc/reference/en/libraries/servo/`
- Arduino Stepper library reference: `https://www.arduino.cc/reference/en/libraries/stepper/`
- The excellent AccelStepper library for advanced stepper motor control: `http://www.airspayce.com/mikem/arduino/AccelStepper/`
- ODrive servo controller: `https://odriverobotics.com/`
- Robotis Dynamixel servo motors: `https://www.robotis.us/dynamixel/`

# 4

# Selecting the Right Arduino Board for Your Project

At the core of every Arduino-based robot is an Arduino board that can be programmed with the Arduino IDE and that runs your Arduino code. In the first three chapters, we exclusively used the Arduino Uno, which is a great board to get started. But there are many more official Arduino boards, as well as third-party boards that are not technically Arduinos but compatible with the Arduino IDE. All of these boards have unique characteristics, strengths, and weaknesses.

In this chapter, we will disentangle the characteristics that matter the most when selecting a board for your Arduino robot. It will be structured as follows:

- Important characteristics of Arduino boards
- An introduction to selected official Arduino boards
- A brief overview of third-party Arduino-compatible boards

## Technical requirements

In contrast to the previous three chapters, this chapter will be mostly informative, without code examples to run.

## Important characteristics of Arduino boards

Arduino boards are simple-to-use yet complex devices with many technical characteristics. It can be difficult to know what all of these characteristics mean, to understand which ones really matter for your project, and what the advantages and disadvantages are that come with them. In this section, we will shed light on the most relevant characteristics of Arduino boards to help you understand what choices you need to make when choosing between boards.

## Input voltage and operating voltage

There are two voltages you need to pay attention to: the board's **input voltage** and its **operating voltage**. The input voltage is the voltage that you can supply to the board's power connector (if it has one) to power it. For example, for the Arduino Uno, the recommended input voltage range is between 7V and 12V. It is capable of handling input voltages between 6V and 20V, but anything outside the recommended voltage range can cause undesirable side effects, such as power instability or overheating. The supplied input voltage is converted by an onboard voltage regulator to the operating voltage.

The operating voltage is what powers the microcontroller, and it determines the voltage of its output signals and its **compatibility** with input signals. This compatibility aspect is why it is so important to keep the operating voltage in mind when selecting your Arduino board. If you use an Arduino board with an operating voltage of 3.3V and connect it to a sensor with 5V output signals, the higher voltage from the sensor can damage your Arduino's microcontroller. If you use an Arduino board that runs on an operating voltage of 5V and connect it to a sensor that uses a 3.3V signal, you run the risk of damaging the sensor with the higher voltage from your Arduino. However, there are boards, sensor modules, and other peripherals that run at an operating voltage of 3.3V but are 5V compatible, meaning that you can safely connect them to a 5V Arduino without the risk of damage.

Because you are much less likely to damage a 5V-compatible Arduino board, 5V boards are considered more robust. Being a 5V board is one of the features that make the Arduino Uno such a robust, beginner-friendly board.

The input voltage is typically less relevant for a robotics project since your robot will likely have a power system that generates the operating voltage(s) for all the components. This way, you can power your Arduino, your sensors, and all other devices with their operating voltages directly without having to use their own internal voltage regulators.

When you select components for your robot, there are three scenarios with regard to their operating voltages you can find yourself in, ordered from most to least preferable:

- You can find versions of all the components you need (Arduino board, sensors, actuators, motor drivers, displays) that have the same operating voltage, either 5V or 3.3V. This is the best case, as it makes your power system simple: you only need to provide one logic voltage. This virtually eliminates the risk of damage from voltage-level mismatches.
- You have some components that run on 5V and some that run on 3.3V, but all their interfaces are 5V compatible. In this case, you need a power system that can provide both 3.3V and 5V, but you can at least safely interconnect all the interfaces of your robot's components without risking any damage.

- Your components run on different operating voltages, and the voltage levels of their interfaces are not compatible with one another. This is not a big problem; it just makes things a little more complicated. In addition to providing the correct operating voltage of either 3.3V or 5V, you will also need to add dedicated level shifters between the incompatible devices you want to connect. It is easy to find breakout boards with several bidirectional level shifters that translate 3.3V signals on one side to 5V signals on the other, but having to add them increases system complexity.

Besides the voltage levels of their interfaces, the type of interfaces that an Arduino board provides is a very important characteristic when choosing a suitable board for your robot.

## Interfaces

Much of what makes Arduino boards so incredibly useful is the ability to easily interface with a very large variety of other components such as sensors, actuators, other Arduinos, or a PC. To this end, having the interfaces you need for your project is often the most important criterion when selecting an Arduino board. Even though you can add additional interfaces to your Arduino Uno with shields and other extension boards, selecting a board that has the required interfaces natively makes for a less complex system overall. Let us look at the kind of interfaces you can choose from when selecting the right Arduino board for your robot, category by category. This section will discuss the interfaces shown in the overview diagram in *Figure 4.1*:

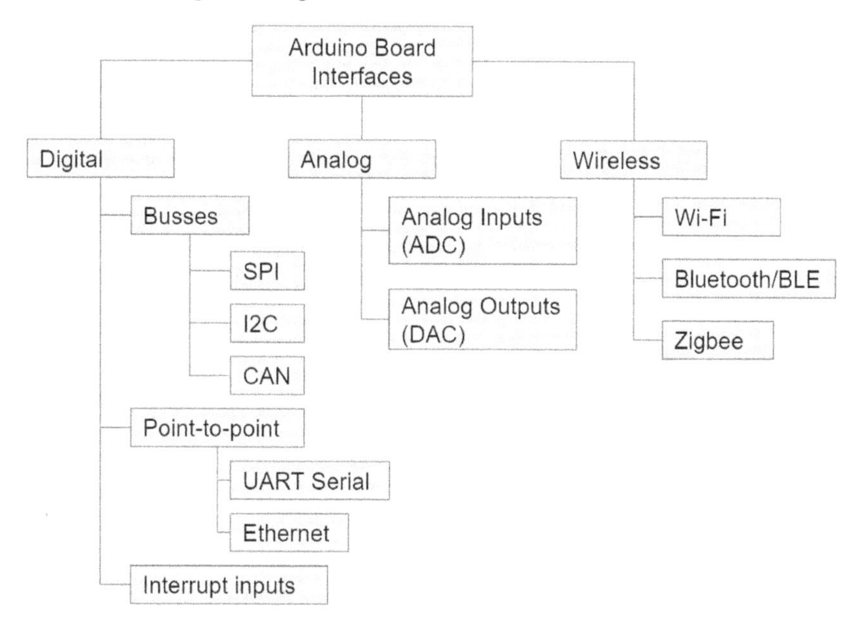


Figure 4.1 – Overview of the Arduino interfaces covered in this section

There are several other interfaces we could discuss, but for the purpose of DIY robots, this selection covers the most relevant ones, and then some.

### Digital interfaces

We have already looked at the most common digital interfaces when we learned about sensor interfaces in *Chapter 2, Making Robots Perceive the World with Sensors*: **Inter-Integrated Circuit (I2C)**, **Serial Peripheral Interface (SPI)**, and the **Universal Asynchronous Receiver/Transmitter (UART)**-based Serial interface. Different Arduino boards differ in the number of these interfaces they provide.

**I2C** and **SPI** are *buses*, which means that a single I2C or SPI interface can connect to multiple targets (such as sensors), and thus one of these interfaces is often sufficient even if your robot has multiple I2C or SPI sensors. For I2C, this comes with the caveat that you need to make sure that all your sensors have unique I2C addresses. If this is not the case, the best solution is to use an Arduino board with multiple I2C interfaces such that you can split your sensors across buses with only unique addresses. Otherwise, you will need to use additional hardware such as I2C splitters (or multiplexers) or *I2C address translators* to circumvent the problem, at the cost of adding additional components and complexity to your project.

In contrast to I2C and SPI, the **UART-based Serial** interface is a **point-to-point (P2P)** connection that can only interface with one peripheral at a time. Each Arduino board has at least one Serial port that is needed for programming. You will want to avoid using this interface for anything else (such as a GPS sensor). Otherwise, you need to disconnect your sensor every time you want to program your robot or use the Arduino IDE's serial monitor. So, whenever your Arduino needs to communicate with a peripheral over UART, you should choose a board that has at least two serial interfaces. While you can use `SoftwareSerial` for initial testing (like we did for our experiment with the GPS receiver), this can have several side effects that are hard to debug, and it is good practice to avoid using software Serial in your robot program.

If you have a board with multiple serial ports, they will be marked with **TX0** and **RX0**, **TX1** and **RX1**, and so on. The familiar calls to serial functions such as this refer to serial port 0:

```
Serial.begin(115200);
```

To use Serial port 1, all you need to do is replace `Serial` with `Serial1` in all calls to the serial interface, like so:

```
Serial1.begin(115200);
```

Digital pins that can serve as **external interrupt inputs** when their voltage level changes from high to low or from low to high are the interface of choice for a variety of sensors. They enable very efficient handling of events that happen at a high frequency, such as decoding **pulse width modulation (PWM)** signals or counting the steps of an incremental encoder. They can also be useful to react incredibly fast (with minimal latency) to any kind of digital input—for example, an emergency stop signal. When you select your Arduino board, you want to make sure it has enough interrupt-capable pins for your needs. It is close to impossible to make up for the lack of interrupt pins without substantial sacrifices in terms of system complexity or code performance.

Especially on the more basic Arduino boards, interrupt pins are a very scarce resource: the Arduino Uno—for example—only has two of them, just enough to sample a single quadrature encoder at full resolution. On more powerful boards with microcontrollers that are based on modern architectures such as the Arduino Due, all digital pins can serve as interrupt inputs.

Other digital interfaces that are available on some Arduino boards that we have not mentioned so far are **Controller-Area-Network** (**CAN**) and **Ethernet**. These are used much more rarely in standalone Arduino robots, but they are supported by the Arduino IDE.

**CAN** is a common bus interface for industrial-grade components such as hydraulic valves or servo motor controllers. To communicate over CAN, these components usually use higher-level application layer protocols on top of raw CAN messages, such as CANOpen or DeviceNet. Native CAN hardware on Arduino boards is not as well supported by the Arduino software tools as the other interfaces, and the CAN application layer protocols are not easy to implement in Arduino code. Furthermore, an additional CAN transceiver is required to connect an Arduino with a CAN interface to a CAN bus. All of this makes CAN an unpopular choice for DIY projects. This interface is instead used mainly for projects geared toward more industrial applications.

**Ethernet** is an extremely powerful and versatile P2P connection, but it is not often used in DIY robots. However, the Arduino IDE comes with the `Ethernet.h` library that lets you use certain Ethernet hardware modules with SPI interfaces right out of the box. While network connectivity can be a powerful feature for a mobile robot, it is usually better implemented over a Wi-Fi interface than over Ethernet. An exception can be the need for reliable, tethered communication over long distances. Applications with this requirement include underwater robots to explore flooded caves or water cavities under ice—not a very common domain for DIY robotics. That being said, wired Ethernet can be extremely useful if your robot is part of a stationary DIY home automation system that you want to control from anywhere.

### Analog interfaces

Analog inputs are great to interface with simple sensors. If you are planning to use analog sensors in your robot, the two things to look out for when selecting your Arduino board are as follows:

- The number of analog inputs (enough for your robot?)
- The resolution of the Arduino's onboard **Analog-to-Digital Converter** (**ADC**)

The Arduino Uno's ADC has a resolution of 10 bits, or 1,024 unique values, as we already learned. If you use a high-quality analog sensor, you might want to be able to sample its output voltage with a higher resolution. Some Arduino boards have 12-bit ADCs (4,096 unique values), and the very high-end Arduino Portenta H7 even sports a 16-bit ADC, resolving an input voltage into 65,536 unique values.

If you want to read an analog value from pin A0 for example, the call to analogRead(A0) defaults to a 10-bit resolution, returning a value between 0 and 1023. If you are using a board with a higher-resolution ADC, you need to set the analog resolution accordingly before calling analogRead() to take advantage of the more capable hardware. If pin A0 is connected to a 12-bit ADC, this would look like so:

```
// Set ADC resolution to 12 Bit.
analogReadResolution(12); int adc_val = analogRead(A0);
```

Keep in mind that using an Arduino with a higher-resolution ADC does not automatically give you a better sensor signal. Using a better ADC only pays off when the signal coming from your sensor is of equally high resolution and has low noise.

Another analog interface that we have only briefly mentioned so far is *true analog output*, enabled by a **Digital-to-Analog Converter** (**DAC**). While we have used the analogWrite() function to command RC servos and modify the speed of DC motors, this function does not actually produce an analog output signal. It produces a very much *digital* PWM signal. The Arduino Uno's microcontroller does not include DAC hardware; however, some other Arduino boards do. The associated pins, labeled **DAC**, can generate an output voltage between GND and the board's operating voltage. This can be a useful feature to drive speakers to generate sound or to interface with analog circuits built from operational amplifiers.

If you use the analogWrite() function with any of the DAC pins of your board, it will generate a true analog output voltage instead of a PWM signal. Similar to modifying the resolution of analogRead() for more capable Arduino boards, you can also modify the resolution of analogWrite() for boards that support higher PWM (or DAC) resolutions than the default 8-bit. For example, to set the output voltage of the DAC1 pin of the Arduino Due with its 12-bit DAC to half the operating voltage (V_o) while using the full 12-bit resolution (4,096 unique values), you would do the following:

```
// Set DAC resolution to 12 Bit and generate 50% V_o.
analogWriteResolution(12);
analogWrite(DAC1, 2047);
```

Analog outputs are an advanced and relatively rarely used feature in Arduino robots, but it is good to keep in mind that there are Arduino boards available that offer this capability should you ever need it. As usual, the Arduino IDE and core library make it easy to use.

### Wireless interfaces

All interfaces that we have discussed so far are wired interfaces, but there is—of course—the class of wireless interfaces too. These are particularly useful for mobile robots that need to maintain contact with a base station or remote control but should not be hindered by a tether that can break, get stuck, or get entangled.

Most microcontrollers do not have any wireless hardware built in (as opposed to I2C and SPI interfaces, for example). Managing wireless communication is typically a complex task that requires dedicated hardware, such as a Wi-Fi module. Your Arduino microcontroller can communicate with this wireless hardware over a wired interface, such as SPI or UART Serial, and the wireless module takes care of the complexities of handling the wireless transmission. The most used wireless interfaces in DIY Arduino robots are Wi-Fi, Bluetooth, and Zigbee. There are official Arduino boards that include Wi-Fi and Bluetooth hardware, and Arduino also makes a Zigbee shield. Let us briefly look at each of these interfaces and scenarios in which you might want to add them to your robot:

- **Wi-Fi** is an extremely powerful and versatile wireless interface that allows you to connect your Arduino robot to a local network and even to the internet. It is an excellent choice when you want to give your robot a browser-based interface, or when you want to enable multiple robots to communicate with each other wirelessly. The Arduino IDE comes with the `WiFi.h` library that works with Wi-Fi-enabled Arduino boards such as the Arduino Uno WiFi, or with the Arduino WiFi Shield.
- **Bluetooth**—and its more modern and versatile version **Bluetooth Low Energy** (**BLE**)—is an excellent choice when you want to create a simple connection between your robot and your smartphone. In its simplest form, you can use a Bluetooth interface as a wireless bridge for your Arduino's serial interface. There are apps that make it easy to create a GUI on your smartphone that lets you send simple commands to your Arduino. Your Arduino program can then parse these commands using the utility functions for the Serial interface. For using many of the more advanced features of BLE on a board that supports it, you can use the `ArduinoBLE` library.
- **Zigbee** devices such as the popular XBee transceivers are a family of versatile wireless modules that enable a range of applications, including P2P connections and mesh networks. What makes them particularly interesting for us is that they can offer a much greater radio range than Wi-Fi or Bluetooth, especially outdoors. In addition, in their most basic configuration, they can act as an easy-to-use wireless bridge for a UART Serial connection, either between two Arduino boards or between your Arduino robot and your PC. You can use Zigbee modules to implement a long-distance radio control or let software on your PC control your robot remotely, or stream telemetry data from your robot back to your PC for real-time visualization. Due to their long-range outdoors, Zigbee modules are great for flying or swimming robots, or any other robots that cover large distances and need to maintain a downlink to a base station.

This brief overview should give you a good idea of the wireless interfaces you can choose from when planning your robot and selecting your Arduino hardware, and you can always refer back to *Figure 4.1* for a general overview of all the interfaces. We will cover all wireless interfaces in much more detail with examples in *Chapter 11, Adding Wireless Interfaces to Your Robot*.

## CPU

Different Arduino boards have different microcontrollers with different **CPU architectures**. The beauty of the Arduino toolchain lies in the fact that we do not have to worry much about these differences.

If it is an Arduino board, we can be sure that the processor architecture, no matter how advanced, is supported by the Arduino IDE. Programming even the most advanced boards will work just as easily as programming a simple Arduino Uno.

Some of the more advanced Arduino features such as CAN and BLE are only available on certain architectures. In these cases, the documentation of the respective library will tell you which board types are supported. You can look at the documentation of the BLE library as an example of this (look for the section titled *Compatibility*): `https://www.arduino.cc/reference/en/libraries/arduinoble/`. When you are planning a more advanced robot project for which you will need a library that only supports certain processor architectures, you need to select an Arduino board that is listed in this library's documentation.

The Arduino IDE does not come with support for all architectures right out of the box since most users will not need them, but you can easily install the necessary files to use any board you want by clicking on **Tools** | **Board** | **Boards Manager...**. In the **Boards Manager**, you can search for the board you want to use and install the required files with one click. After this step, you will find your board in the **Tools** | **Board** drop-down menu.

You can also quickly open the **Boards Manager** by clicking on the board icon on the left side of the IDE, shown in *Figure 4.2*:

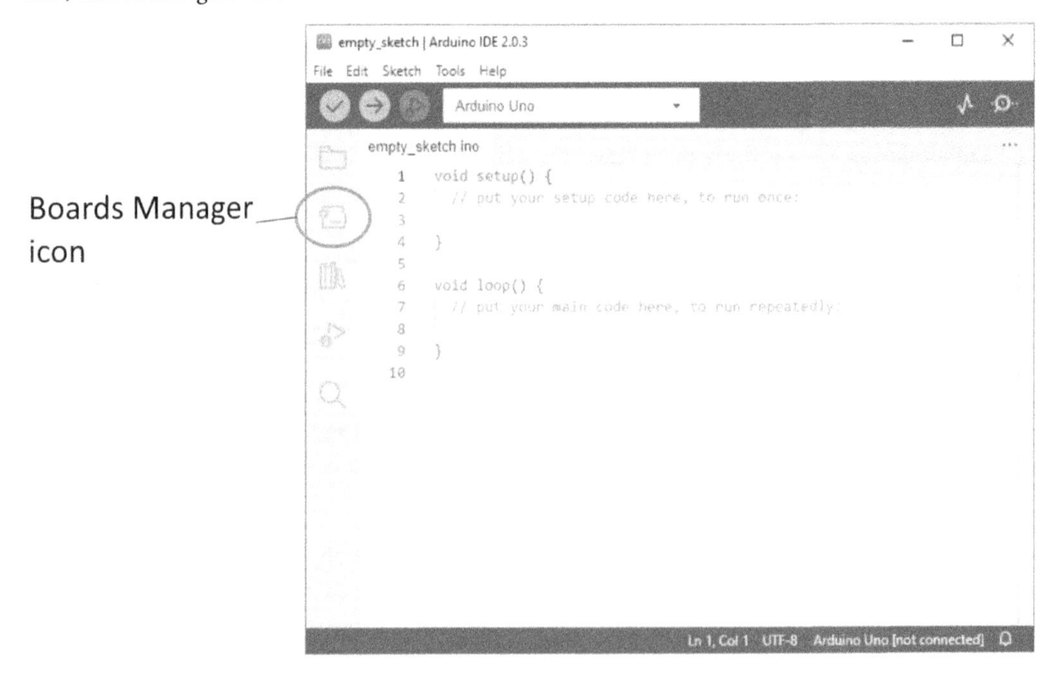


Figure 4.2 – Shortcut to the Boards Manager

Apart from the specific architecture, the two most important characteristics of an Arduino's CPU are its bit width or **register size** and its **clock speed**. Together, they determine how fast the CPU can run your program.

### Register size

A CPU does some amount of computation every clock cycle. The instruction it executes and the data it operates on are stored in the CPU's registers. Registers are small blocks of memory that the CPU can access extremely fast. The larger the CPU's registers, the more work it can do per cycle. The earliest practical microcontroller CPUs, and many that are still in use today, have a register size of 8 bits. This is small and makes them relatively slow. It also means that operations on a 16-bit data type such as an Arduino `int` type take more than one clock cycle, and this can lead to trouble when the value of that integer changes unexpectedly between CPU cycles—for example, due to an interrupt execution. The ATmega328P on your Arduino Uno is an example of an 8-bit processor that is still perfectly adequate for many microcontroller applications.

More capable Arduino boards such as the Arduino Due have 32-bit CPU architectures that can do significantly more in one processor cycle.

### Clock rate

The CPU clock rate determines how many operations the CPU will perform per second. Most Arduino boards are clocked at 16 MHz. This means they can perform 16 million CPU operations per second (which sounds like a lot but is very slow compared to modern PC CPUs that clock in the GHz range, performing billions of operations per second).

A higher clock rate allows you to run more advanced programs on an Arduino board. Processing image data, fast sensor fusion, advanced signal processing of multiple sensors, or any other application that involves complex calculations with floating-point numbers might require a fast CPU.

## Memory

Most Arduino boards have three types of memory: Flash, **Static Random Access Memory** (**SRAM**), and **Electronically Erasable Programmable Read-Only Memory** (**EEPROM**). *Figure 4.3* shows an overview of these three types of memory that are connected to the CPU with its registers:

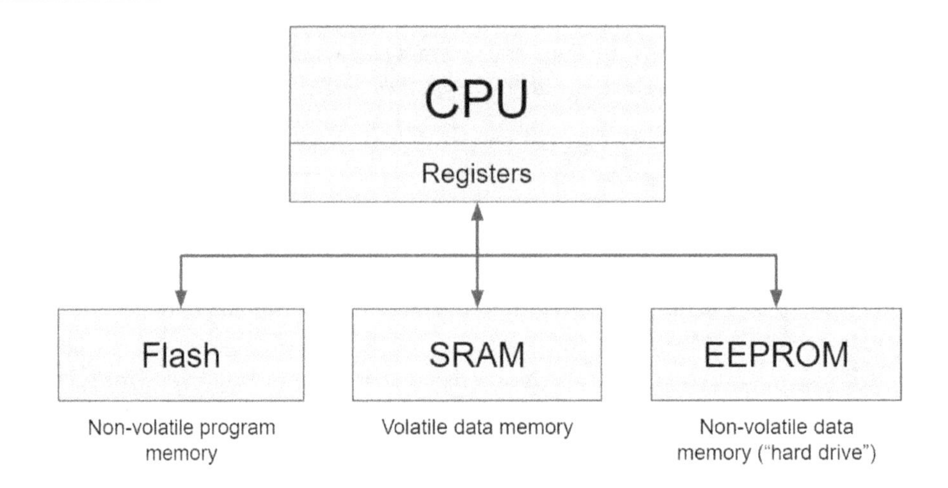


Figure 4.3 – The three main types of memory inside an Arduino microcontroller

Let us look at what each of these is used for and what it means for selecting the right Arduino board for your robot.

### *Flash*

**Flash memory** is non-volatile memory. Its contents will persist across power cycles. Your Arduino uses its Flash memory primarily to store the program that you write for it in the Arduino IDE. In addition, the Flash memory stores the Arduino bootloader that allows you to program the board via USB instead of having to use the **in-circuit serial programming** (**ICSP**) interface each time. Often, programs that we write for DIY Arduino robots are small (in the sense that the compiler turns them into small binaries), and thus we can make do with a small Flash memory. The Arduino Uno—for example—has a small Flash memory of only 32 KB, and that is often enough. But if your programs get bigger or you use many external libraries in your program, you might reach the limits of your Uno's Flash memory. In this case, the IDE will warn you, or if your program is too big, the compilation will fail altogether. The solution to this is to either optimize your code or switch to a board with more Flash, such as the Arduino Mega 2560, which has 256 KB of Flash memory.

### *SRAM*

**SRAM** is where your Arduino creates and maintains variables that you declare and use in your program. SRAM is volatile memory, meaning that its contents are deleted after a system power cycle. When you power on your Arduino, it is safest to assume that the content of the SRAM is random.

> **Note**
>
> The random initial content of SRAM is why it is good coding practice to always initialize variables rather than assuming a certain value after declaration.

A small SRAM can become problematic when you use many large variables in your program. It is quite common to run out of SRAM on an Arduino Uno (that only has 2 KB of it) due to the use of long `String` variables, such as help texts that you want your code to print on the Serial monitor. Another common reason to run out of SRAM is the use of lookup tables that you might want to speed up certain calculations. Running out of SRAM can make your program behave unpredictably, which is very undesirable and hard to debug. If, during compilation, the IDE tells you that you are nearing the limits of your board's SRAM, you can try reducing the use of large variables (for example, by shortening strings) or switching to a different Arduino board with more SRAM. The Arduino Mega 2560—for example—has 8 KB of it, four times as much as the Arduino Uno. Another workaround is to use the `PROGMEM` keyword to instruct the compiler to store certain `constant` variables in Flash memory. This is an advanced technique that you only need to consider if all your other options are exhausted. You can learn more about it and see the example code here: `https://www.arduino.cc/reference/en/language/variables/utilities/progmem/`.

### *EEPROM*

**EEPROM** is a non-volatile memory that you can use to store parameters or any other data between power cycles, even for extended periods of time. It is like a tiny hard drive within your Arduino. The Arduino Uno has 1 KB of EEPROM, and you can easily write to and read from it with the `EEPROM` library. The difficulty of using data from EEPROM lies in ensuring that it is valid. If you tell your robot to use parameters that are at a certain EEPROM address, but, for any reason, the data at this EEPROM address has been overwritten, unintentionally altered, never initialized, or deleted, your robot could behave in unintended ways because of that. One way to check the EEPROM content for validity is to calculate and maintain an EEPROM **checksum** or **cyclic redundancy check** (**CRC**).

## Additional hardware

Some Arduino boards come with additional hardware besides the microcontroller and interface modules that can make building your robot easier. When you have specific hardware in mind that you will need for your robot, such as an **inertial measurement unit** (**IMU**) or temperature sensor, it is always worth looking for an Arduino board that might already have it integrated to simplify your design and spare you the cost of additional hardware. More examples of such additions are real-time clocks to keep accurate track of time even when the Arduino loses power, battery management systems for an attached lithium battery, or SD card readers. A battery management system automatically charges the battery when the Arduino is powered externally and runs the board from the battery when external power is disconnected. An SD card reader can be incredibly useful to save substantial amounts of logging data that your robot might collect on its mission (for example, as GPS data) for later analysis. The Arduino IDE comes with the `SD` library, which makes it easy to write to and read from SD cards via the Arduino's SPI interface.

## Size

The physical size of your robot design might constrain the size budget that you have for your Arduino board. Fortunately, Arduino boards come in a variety of form factors, and especially some third-party Arduino-compatible boards are incredibly small. It is preferable though to use a bigger over a smaller board whenever possible, as it is usually easier to work with the bigger boards. You might have noticed that there is a lot of unused space on your Arduino Uno, and it would be easy to squeeze the same functionality into a much smaller board. The reason it is so big is simply for better usability. The Arduino Nano packs almost the same functionality as the Arduino Uno into roughly half the board size.

If you are constrained in terms of board size but not so much in height, you can also consider using an Arduino Uno with shields for additional functionality such as Ethernet or SD card access. However, it is best to not use more than one shield in each application. Shields often use the same pins and might unexpectedly interfere with each other's functionality.

## Cost

One of the factors that make the Arduino ecosystem so attractive and accessible is the low cost of the hardware. However, not all Arduino boards are equally cost-effective. At the time of writing, the Arduino Nano Every is the most affordable board in the official Arduino store for a little more than $10. This makes it especially great for school projects or when you want to build a swarm of identical robots. On the other side of the spectrum lie very powerful Arduino boards such as the Portenta H7, which can cost upward of $100.

If you are looking for an affordable board, it is always worth considering sources outside of the official Arduino store. Because many of Arduino's designs are open source, there are many great alternatives that are virtually identical but cost a fraction of the original product. Support through the IDE might not be quite as good, and you might need to install additional drivers to account for different (more cost-effective) USB-to-Serial hardware, but it can be worth the trouble to save on one of the more expensive components of your robot.

# An introduction to selected official Arduino boards

The list of official Arduino boards is long and growing. At this point, you know what the significant differences between the boards are and what to look for when you select an Arduino board for your next robot project. Feel free to put your new knowledge to the test and see if you can narrow down your choice for a board that meets your requirements the best from the list of Arduino boards here: `https://store-usa.arduino.cc/collections/boards`.

Let us also look at two direct alternatives to the Arduino Uno that you are likely to want to consider at some point: the Arduino Nano and the Arduino Mega 2560.

## Arduino Nano and Nano Every

Especially when you build a very small or lightweight robot, you will quickly find the Uno's form factor a bit bulky. If otherwise the features and performance of the Arduino Uno are sufficient, the Arduino Nano can be the perfect alternative (see *Figure 4.4*). Just as with the Arduino Uno, it uses an ATmega328 microcontroller, only in a surface-mounted package to save space. This means it has the same amount of Flash and SRAM memory, and every program you wrote for the Arduino Uno will run just as well on the much smaller Nano. The Nano has the same indicator LEDs and it even has a voltage regulator on the bottom side of the board, which means just as with the Arduino Uno you can run it with input voltages between 7V and 12V. It uses a USB Mini-B jack instead of the Uno's much larger USB B jack to further reduce size. The biggest difference in terms of usability is that the Arduino Nano does not have female pin headers. You can add them yourself, or you can add male pin headers to mount them on a solderless breadboard for prototyping:

Figure 4.4 – Top side of the Arduino Nano, the "little brother" of the Arduino Uno; it is designed to be compatible with a solderless breadboard

If you are comfortable moving away from the familiar ATmega328 microcontroller, it is worth considering the Arduino Nano's latest evolution, the Arduino Nano Every. This board features the ATMega4809 microcontroller and is better in almost every aspect, offering a higher clock speed and more memory, while also being cheaper.

## Arduino Mega

If, instead of the size of the Arduino Uno, the problem is a lack of memory or interfaces, the Arduino Mega 2560 Rev3 might be the ideal alternative, shown in *Figure 4.5*. In some sense, it is the bigger brother of the Arduino Uno. The form factor and pinout of one side of the board are identical to the Arduino Uno, making it a perfect drop-in replacement. Most of the code you wrote for your Arduino

Uno will work without any changes on the Arduino Mega 2560. Even the way you connect peripherals to it stays mostly the same. Because the pinout is almost identical to that of the Uno, all the Arduino shields that were initially designed for the Arduino Uno also fit onto the Arduino Mega 2560. The difference between the boards is that the Arduino Mega 2560 extends further than the Uno, giving you an additional eight analog input pins, three more UART serial ports, an additional I2C bus interface, and 32 more GPIO pins. 15 of its pins are capable of PWM output (compared to six on the Uno) and it has six hardware interrupt pins, three times as many as the Uno. The Mega 2560 also has eight times the Flash memory, four times the SRAM, and four times the EEPROM capacity of the Uno, making it a substantial upgrade with regard to memory:

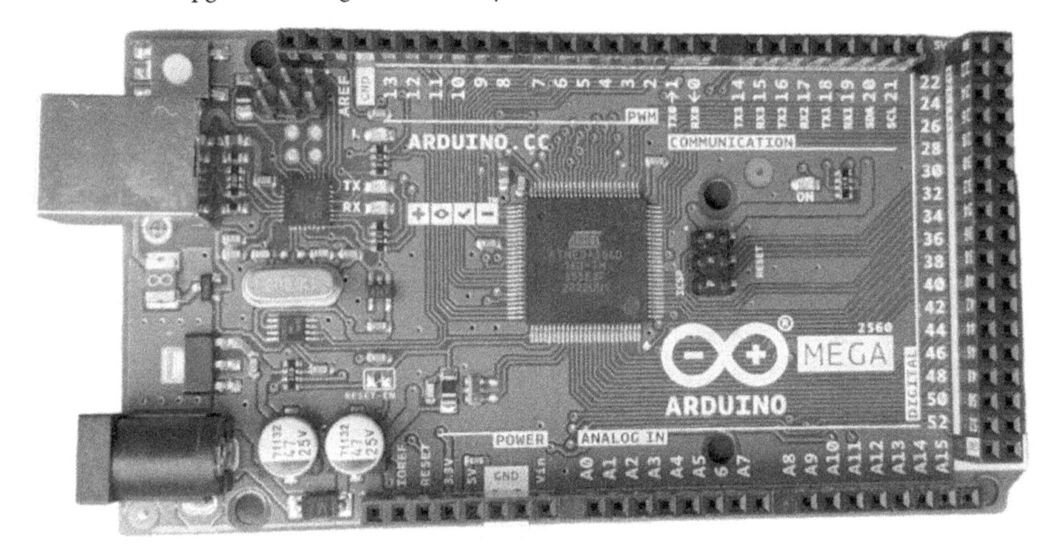


Figure 4.5 – An Arduino Mega 2560, the "big brother" of the Arduino Uno; the dimensions and pinout of the left half of this board are identical to that of the Arduino Uno

The Arduino Mega 2560 achieves this performance gain by using a surface-mounted version of the ATmega2560 microcontroller that contains the same CPU as the Uno's ATmega328P (8-bit, 16 MHz) but packs a lot more peripherals, memory, and GPIO.

## A brief overview of third-party Arduino-compatible boards

Because the Arduino ecosystem is so popular, there are many manufacturers of alternative boards that fill special niches that none of the official Arduino boards fills. Often, that means that these boards are cheaper, have special hardware, or use more powerful processors. The variety is large, and the space is quite dynamic, which makes it hard to curate a complete list of boards that you can use with the

Arduino toolchain. But here is a small list to get you started in your search for alternative boards that might be just the right fit for your robot.

## Adafruit Feather and Trinket M0

Adafruit is a maker and seller of a large variety of breakout boards for interesting components, and not only are their boards usually very well done, but they are also exceptionally well documented, complete with open source libraries and tutorials. Besides breakout boards for small components such as sensors, they also make development boards that you can program with the Arduino IDE. Their line of *Feather* boards uses a very small form factor and is made for use with a solderless breadboard, similar to the Arduino Nano. But many of the Feather boards feature processors that are an order of magnitude more powerful than the ATmega chips you find on a typical Arduino, yet they do not cost more. They also come in versions with all kinds of interesting hardware (Wi-Fi, BLE, SD card) and are an excellent choice when you are looking for a small, affordable, and powerful board (`https://learn.adafruit.com/adafruit-feather`). *Figure 4.6* shows a Feather board with a very capable CPU, as well as Wi-Fi and Bluetooth interfaces next to an Arduino Uno:

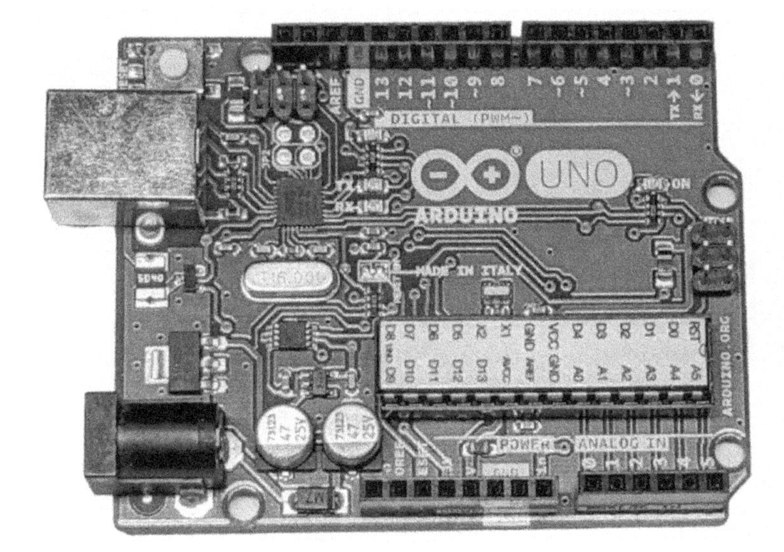

Figure 4.6 – A classic Arduino Uno next to a much more powerful board with integrated Bluetooth and Wi-Fi hardware in the Adafruit Feather form factor

Adafruit also makes the Trinket M0, a tiny board with few pins, but an immensely powerful 32-bit, 48 MHz processor with plenty of memory to run even advanced robot programs (`https://www.adafruit.com/product/3500`).

Many Feather boards as well as the Trinket M0 can even run the CircuitPython interpreter to support the use of the Python programming language.

## SparkFun RedBoard

Another big-name maker and seller of Arduino accessories and general electronics components is SparkFun. SparkFun, too, sells its own take on the Arduino board: the SparkFun RedBoard lineup (`https://www.sparkfun.com/redboards`). Almost all of them are based on the physical dimensions and pinout of the Arduino Uno, but they come with a range of additional extra features, more powerful processors, and additional interface connectors.

## Teensy

The boards of the Teensy line by PJRC are small Arduino-compatible boards with very powerful microcontrollers. The latest generation at the time of writing, the Teensy 4, uses the high-performance Cortex-M7 CPU, a 32-bit machine running at a clock rate of 600 MHz, close to 40 times faster than the Arduino Uno. Many Teensy boards have multiple SPI, IC2, Serial, and CAN interfaces, as well as true analog outputs and native SD card support. As with the Arduino Nano and the Adafruit Feather boards, Teensy boards do not have female headers (even though you can always install them yourself) and are instead meant to be used in combination with a solderless breadboard. You can see an example in *Figure 4.7*:

Figure 4.7 – The deceptively small yet immensely capable Teensy 3.6, with a micro USB jack on the left and a micro SD card slot on the right

Teensy boards run at 3.3V, so you will need to take the appropriate precautions when you want to combine them with 5V hardware. To integrate Teensy into the Arduino toolchain, PJRC provides **Teensyduino**, a free add-on to the Arduino IDE that makes programming Teensy boards seamless.

## Summary

Once you start planning your Arduino robot, there are a lot of boards to choose from—both original Arduino boards and many more third-party, Arduino-compatible boards. After reading this chapter, you know what to look for when you choose your perfect board: operating voltage, interfaces, CPU, memory, additional hardware, size, and cost. You know what the key characteristics in these categories are, the trade-offs, and how to work around some of the limitations if you cannot find the perfect board. Have fun using your newly gained knowledge and explore the world of boards for your DIY Arduino robots!

This chapter concludes the first part of this book, in which we took an in-depth look at the core hardware components of your robot. Hopefully, you feel inspired to get your hands on some of these components and start building a robot! But before we start building physical robots, the next part of the book will teach you the fundamentals, advanced techniques, and best practices of how to write the software for your robot. We have a lot to cover in the next chapter, so let us get to it.

## Further reading

The following resources contain many valuable pointers that let you explore this chapter's topic further:

- A guide to selecting microcontrollers. Also, an example of the excellent educational content Adafruit produces: `https://learn.adafruit.com/how-to-choose-a-microcontroller`.
- A non-exhaustive but informative list of Arduino-type boards on Wikipedia: `https://en.wikipedia.org/wiki/List_of_Arduino_boards_and_compatible_systems`.

# Part 2: Writing Effective and Reliable Robot Programs for Arduino

This part teaches you about the essential concepts and techniques needed to write effective and reliable programs for Arduino robots efficiently, taking into account the computational limitations of microcontrollers. In addition, it introduces you to object-oriented programming with C++ and walks you through the process of creating your own object-oriented, reusable Arduino libraries. This part also includes the best practices for testing and debugging with the Arduino IDE's powerful serial interface tools: the Serial Monitor, and the Serial Plotter.

This part has the following chapters:

# 5

# Getting Started with Robot Programming

We have already used a few small code snippets here and there to demonstrate certain functionalities of the Arduino Uno to read sensors and control motors. But what if you wanted to combine all of this into a single program that uses multiple interfaces at the same time, runs control algorithms, and controls motors, all concurrently?

In this chapter, you will learn how to make all of this work in practice. It is structured as follows:

- The general structure of a robot program
- Cooperative multitasking
- State machines
- Blocking and unblocking function calls
- Feedback control

You will be able to use these general concepts and techniques in every robotics project.

## Technical requirements

There are several code examples in this chapter to illustrate the concepts that we cover. You can run all of these code examples on your Arduino Uno. We will also cover feedback control, and as a practical example, we will develop the code to turn a DC motor with an encoder into a position-controlled servo motor. For this example, you will need your Arduino Uno, a DC motor with a quadrature encoder, and a motor driver PCB, ideally with an ST L298 driver chip (since this is what we use in the example). The Arduino motor shield will not work for this example unless you are comfortable modifying it to reroute some of its wire traces.

## The general structure of an Arduino robot program

We have already gotten to know the general structure of an Arduino program, consisting of three main parts:

- `#include` statements and declarations of global variables at the top.
- The `setup()` function with all the necessary one-time initializations. It is called automatically at the start of the program execution.
- The `loop()` function, which is executed in an infinite loop right after the `setup()` function has finished.

We can define many more functions that we can use inside the `setup()` and the `loop()` functions. This helps to keep these two main functions concise and readable, and it makes our program easier to maintain and modify. We usually write these functions at the end of the program, underneath the `loop()` function. An example of this is the `mcp_read_channel()` function that we developed in *Chapter 2, Making Robots Perceive the World with Sensors*.

This general structure is the same for every Arduino program, and robot programs are no exception. But there are many different ways to implement the control logic inside the `loop()` function, and robot programs have some specific requirements that we need to take into account.

Most importantly, our Arduino CPU has to work on multiple tasks concurrently. It will usually sample a variety of sensors over various interfaces, perform sensor signal processing, print output over the serial port, react to commands from the serial port, blink LEDs, run control algorithms, and control actuators. Some of these **tasks**—for example, feedback control loops—need to be executed at very reliable time intervals. If we don't update the control outputs at the right interval, our system might become unstable in the control-theoretical sense. The joints of a robotic arm, for example, might start vibrating or even shaking, or a Segway-style balancing robot might fall over. A robot operates in the real world, and the laws of physics are unforgiving. Physics will not wait for our controller to update if it takes longer than it should. It is this interaction with the real world, governed by the unforgiving laws of physics, that sets robot programs apart from most other programs. While speed and reliable timing are mostly nice-to-have features for user-facing programs, they can be existential for a program that controls a robot. The ability to guarantee very reliable timing in the execution of the program is called **real-time** capability, and many systems that physically interact with the real world need to be real-time capable.

To ensure real-time capability, all of the different tasks that run on your Arduino inside the `loop()` function need to be cooperative: none of them is allowed to occupy the CPU for an extended period of time in order to not delay the execution of other tasks. On more advanced robots that have CPUs with multiple cores, it is common to dedicate at least one core to the tasks that require real-time capability and run non-real-time tasks on the other cores. But most Arduino microcontrollers only have one core, so all tasks need to be implemented in a real-time capable manner. An example of a task that is not cooperative is the blink program that we wrote in *Chapter 1, Introducing Robotics and the Arduino Ecosystem*. In between switching the LED on and off, we simply called `delay(500)` to let the CPU idle for 500 **milliseconds** (**ms**). No other task could run in these 500 ms, and the entire CPU capacity is wasted on waiting. You will not see calls to `delay()` in real-time robot programs. In a cooperative implementation, the blink task would change the LED state when it needs to, which takes a very small amount of CPU time, and then immediately release the CPU for other tasks to use.

Fortunately for us, it is easy to implement a real-time system on a microcontroller—much simpler than it would be on a PC, where the **operating system** (**OS**) has the final say about when tasks are executed. On a microcontroller without an OS, we have full control over the execution of all tasks and can easily use that to our advantage.

# Cooperative multitasking

Multitasking describes a mechanism that lets a single CPU core serve multiple tasks concurrently. There are multiple ways to implement multitasking. A key distinction can be made between **preemptive multitasking** and **cooperative multitasking**. In preemptive multitasking, a higher-level authority such as the OS decides how long a given task can use the system's resources (such as the CPU) before they are given to another task. This is especially useful when a lot of different tasks run on a large system, and not all of them can be trusted to manage their resource consumption responsibly. However, it makes the system more complex by requiring a powerful task manager that can start and stop task executions.

In cooperative multitasking, tasks are split into small steps and then served sequentially—for example, in the Arduino `loop()` function. Each task only does a small amount of computation at once, requiring only a very short usage of the system's resources. It then automatically releases the CPU for the next task. All tasks are implemented such that even in the worst case, all tasks can be executed once within the designed cycle time of the loop frequency—for example, 1 ms. This is great for simple systems (such as microcontrollers) because it does not require an OS or another task manager. However, implementing tasks such that they do their work in small pieces each time they are called is not always easy. But not to worry—we will cover all the necessary techniques in this chapter.

## Running two tasks asynchronously

Let us get right to an example of the abstract idea of cooperative multitasking. In this example, we want to execute two tasks concurrently without them disturbing one another. We want to blink an LED, and we want to periodically print some serial output, and we want these two tasks (the blink task and the print task) to be independent of each other. To be able to add many more tasks later, we need to avoid the use of `delay()` at all costs. The following code represents an example implementation of this. Take a look at it, and then we will discuss it in detail:

```
// Variables of the blink task.
unsigned long last_blink_time;
int blink_interval = 200;
```

```
// Variables of the print task.
unsigned long last_print_time;
int print_interval = 1000;

void setup() {
  Serial.begin(115200);
  pinMode(LED_BUILTIN, OUTPUT);
  // Initialize last_time variables.
  last_blink_time = millis();
  last_print_time = millis();
}

void loop() {
  // Check if it is time to run the blink task.
  if (millis() - last_blink_time >= blink_interval) {
    // Update the last blink time.
    last_blink_time += blink_interval;
    // Execute the blink task.
    digitalWrite(LED_BUILTIN, !digitalRead(LED_BUILTIN));
  }

  // Check if it is time to run the print task.
  if (millis() - last_print_time >= print_interval) {
    // Update the last print time.
    last_print_time += print_interval;
    // Execute the print task.
    Serial.println("Tick");
  }
}
```

Let us focus on the content of the `loop()` function. Two blocks of code cooperatively manage two tasks: blinking the built-in LED and periodically printing the word `"Tick"` to the Serial Monitor. These two code blocks look very similar to one another. They consist of a few lines that control the task execution, and the task itself. The key and what makes these two tasks cooperative is that each line of code, including the actual task, executes extremely fast, which allows the `loop()` function to run at a very high frequency. None of this code will occupy the CPU for a significant amount of time and make the rest of the code wait for it, as a call to `delay()` would.

At the beginning of each block, an `if` statement checks if the desired time interval between consecutive task executions has elapsed. The way this check is implemented in the example might look a little unintuitive to you. The advantage of this particular implementation is that it continues to work fine even if `millis()` overflows, which means if the number of milliseconds since power-on exceeds what an unsigned 32-bit integer can hold. In this case, the output of `millis()` will start at `0` again. Some other, more intuitive ways of writing this timing check might result in undesired behavior in this case.

If the desired interval between task executions has not passed since the last, the program can simply move on to check if it is time for the next task. However, if the timing check evaluates to `true`, it is time to execute the task. But first, the code increments the variable that stores the last execution time by the desired interval between task executions. This guarantees that the task executes again after the next interval has passed, and even if it executes a little late for any reason, these timing errors do not accumulate, and on average, the execution frequency will remain exactly on target. After this little bit of bookkeeping, the actual task is executed. Both of the tasks in this example—the blink task and the print task—are just one line of code that executes very quickly.

At the beginning of this example, we declare the required variables to hold the task intervals and last execution timestamps (in milliseconds), and we initialize these timestamps in the `setup()` function to the current time.

One of the advantages of this framework is that we can very easily add many more tasks to this list. As long as each of them only contains small statements that execute fast, they will not interfere with one another.

## Task execution with variable frequency

To demonstrate the independence of our two tasks, let us make the blinking frequency variable. With one very small addition, we can control the blinking frequency with a potentiometer without affecting the frequency of the print task. No matter how fast or slow the LED blinks, the print task will stoically print the word `"Tick"` every second. Try to think about how you would implement this change, and try it yourself before you continue to read.

All we need to do is add this line at the top of the `loop()` function:

```
blink_interval = analogRead(A0);
```

This line also executes extremely fast and does not cause any delay, so it fits well into the cooperative multitasking scheme. If you add a potentiometer to your setup and connect its output pin to the Arduino's `A0` analog input, you can control the time between switching the LED on or off. The controllable range with the potentiometer is from 0 ms, which means it will flicker as fast as possible to `1023` ms or roughly 1 second. Notably, if you look at the Serial Monitor output, the print task is not affected by this change. This highlights how easy cooperative multitasking makes it to run multiple tasks asynchronously and independently from one another.

## State machines

In our introductory example of cooperative multitasking, we chose two tasks that are extremely small and that do the exact same thing each time they are executed. However, tasks can get much more complicated and might require us to execute several different steps each time they are called. In other words, we need to break a big chunk of code up into small pieces. A great way to implement these more complex tasks in the cooperative multitasking framework is by writing them as **state machines**.

A state machine is a piece of code that keeps track of the **state** of the task (hence the name). Certain conditions, such as timing checks, can trigger a transition between states, which causes a bit of code to be executed. The LED blink task from the last example can easily be implemented as a state machine. This adds a few lines of code, but it makes it much more readable. In the blink task state machine, the state is the state of the LED, which is either ON or OFF. The transition between these two states is naturally implemented as turning the LED on or off, respectively. To implement this, we can start with the previous example. Right before the `setup` function, we add a custom `enum` object with the two states, and a variable to hold the LED state:

```
// The states of the blink task.
enum state {
  LED_OFF,
  LED_ON
};
// The blink state variable.
uint8_t led_state = LED_OFF;
```

The states are really just numbers, but to make the code more readable and to highlight the logical meaning of the states, it is good practice to give them names with the help of an `enum` object. This allows us to use the more meaningful labels `LED_OFF` and `LED_ON` instead of the numbers `0` and `1`.

Let us modify the block inside the `loop()` function that controls and executes the blink task to look like this:

```
if (millis() - last_blink_time >= blink_interval) {
  // Update the last blink time.
  last_blink_time += blink_interval;
  // Execute the blink task.
  blink_task();
}
```

Now, we need to implement the blink state machine inside the `blink_task()` function at the end of our program, and that is where it gets interesting! Take a look at this blink state machine:

```
void blink_task() {
  switch (led_state) {
    case LED_OFF:
      // LED is off, we need to turn it on.
      digitalWrite(LED_BUILTIN, HIGH);
      // Update the state variable.
      led_state = LED_ON;
      // Update the blink interval (just for fun).
      blink_interval = 50;
      break;
    case LED_ON:
      digitalWrite(LED_BUILTIN, LOW);
      led_state = LED_OFF;
      blink_interval = 450;
      break;
  }
}
```

It is very common and practical to implement state machines as a `switch()` statement that selects the appropriate action based on the current state. If the `blink_task()` function is called when the state is `LED_OFF`, the `switch()` statement selects the first case. This block turns the LED on to transition from the `LED_OFF` state to the `LED_ON` state and updates the `led_state` variable accordingly. Something very similar happens when the `blink_task()` function is called when the state is `LED_ON`.

> **Attention**
>
> It is important to end each `case` block with a `break` statement. Forgetting the `break` at the end of a `case` block is a very common bug that can lead to unexpected and confusing behavior of your state machine. Since a `case` without a `break` statement is valid C++ syntax, the compiler will not even warn you.

In each of the two cases, we also update the blink interval. This is not necessary, but it exemplifies how writing the blink task as a state machine gives us more flexibility than writing it in the previous example. With the state machine implementation, we can do different things in each of the two state transitions, and one way to use this capability is to create a blink pattern in which the ON and OFF durations are different. In our example, the LED will be ON for only 50 ms but OFF for 450 ms. With very little change, you could again use the potentiometer to control the ratio between LED on and off time.

To solidify your understanding of state machines, you can try to implement the print task as a state machine, too. Instead of only printing `"Tick"`, you can let it print the words `"Tick"` and `"Tock"` alternatingly, or your first and last name, or `"State"` and `"Machine"`.

## Blocking and non-blocking function calls

To successfully implement cooperative multitasking, each line of code needs to execute extremely fast and not tie up the CPU for any significant amount of time. A slightly more technical way of stating this is to say that all the function calls inside the cooperative multitasking framework need to be **non-blocking**, as opposed to blocking.

A non-blocking function will always do its best to return immediately, and the time it takes for it to return is going to be predictable and very short. Nothing inside a non-blocking function needs to wait for something else. The `blink_task()` function in the last example is a good sample of a non-blocking function. Nothing in it needs to wait, and we can be confident that the function returns virtually immediately.

In contrast, a blocking function will start a process that can take an (unpredictable) amount of time. Examples of this are using the `delay()` function or waiting to receive a certain number of bytes over the Serial port. The blocking function will not return before this process is completed, thereby blocking the overall program execution for the waiting period.

> **Important note**
>
> We always try to avoid using blocking function calls inside the `loop()` function of robot programs. They can introduce unexpected delays and can cause non-repeatable behavior that is extremely difficult and time-consuming to debug.

As we said before, the physical world can be an unforgiving place for robots, with delays or timing inconsistencies in their control algorithms.

If you use the built-in functions provided by the Arduino IDE, it is not always obvious whether the function is blocking or non-blocking. The most likely candidates for blocking function calls are those that involve communication methods. An example of a built-in blocking function is `Serial.parseInt()`. We can use this function to interpret an **American Standard Code for Information Interchange** (**ASCII**) character string received over the serial interface as an integer. It will read characters from the serial input buffer until it encounters the first non-numeric character, at which point it returns the value of the integer represented by the received characters. If the characters were already received and are stored in the input buffer, the function returns immediately. However, if there are no characters in the input buffer by the time the function is called, it will wait until they are received. This can take an unpredictable amount of time. Until then, the function will block the entire program execution. Some blocking functions, including `Serial.parseInt()`, have timeouts to put an upper bound on the time they can block. A **timeout** is a period after which the function gives up and returns a default value or an error code.

When we implement functions ourselves, we should always try to implement them in a non-blocking way. That often results in a more complicated implementation, but the benefit of predictability and compliance with the cooperative multitasking framework is usually worth it. Consider, for example, two versions of a simple `blink()` function to blink an LED. A blocking implementation could look like this:

```
void blink(int led_pin, int on_interval, int off_interval) {
  digitalWrite(led_pin, HIGH);
  // Wait and block the entire program.
  delay(on_interval);
  digitalWrite(led_pin, LOW);
  // Wait and block the entire program.
  delay(off_interval);
}
```

This is a remarkably simple function, and if we call this function periodically inside the `loop()` function, we will see the LED blink, but the CPU will be completely blocked, even though turning the LED on and off every so often is a negligible amount of work. There is no other task the CPU can work on between turning the LED on and off.

On the other hand, we can implement the `blink()` function as in the state machine example, and only call it when it is time to change the LED state. While this is a little more complex, the function call always returns immediately, and it allows the CPU to work on other useful tasks in between switching the LED on and off.

In practice, implementing it as a state machine is often the best method for turning a blocking function into a non-blocking one. It is a common workflow to develop a prototype of a function in a blocking manner, experimenting with different delay values until it works as expected, and then turning this blocking function into a non-blocking version, usually by implementing it as a state machine. This allows you to use it as part of a complex program without interfering with the timing requirements of other parts of the program.

You can often use the blocking functions of the Arduino IDE in a non-blocking manner if you ensure that the conditions for it to return immediately are met. In the case of the `Serial.parseInt()` function, we can safely use it if we know that the serial input buffer contains characters forming an integer and a non-numeric termination character. If we know, for example, that we can expect a double-digit integer and a termination character (such as a *newline*), we can safely use `Serial.parseInt()` without risking that it might block for a long time if there are three characters in the buffer. We can use the `Serial.available()` function to quickly check how many characters are in the Serial input buffer, and if this number is greater than `3`, `Serial.parsInt()` will not need to wait for input. It will not block our program, even though it can be a blocking function. A simple implementation of such a safeguard around the blocking call to `Serial.parseInt()` could look like this:

```
void updateDoubleDigitIntegerIfPossible(int *val) {
  if(Serial.available() == 3) {
    // There are three characters in the buffer.
    // Update the value of val.
    *val = Serial.parseInt();
  }
}
```

This function will only attempt to parse an integer and update the value of the parameter referenced by `val` if there are three characters available in the input buffer. And if we can be sure that these three characters contain two digits and a termination character, it will not block and return immediately.

# Feedback control

Feedback control is another aspect of robotics that is uncommon in other domains, but it is often at the core of a robot program. This section gives you a concise introduction to feedback control, and we will develop a servo motor as an example application. The field of feedback control is vast, and there are many great and math-heavy textbooks on the topic, exploring many fascinating aspects of it. So, while this section can only scratch the very surface, it will give you all the required foundations to use feedback control in your Arduino robot projects.

## What is feedback control?

Feedback control is a technique to control the state of a **dynamical system**. A dynamical system is something that changes its **output** over time under the influence of its internal state and certain **inputs**. Let us consider the example of a servo motor with a position control interface. This servo motor needs a feedback control system to enable its position control capability. Its output is the output position, and it has a sensor that measures it. We cannot simply set the position of an electric motor; all we can do is modify the amount of current that flows through it by modulating the duty cycle of the PWM signal that controls the transistors that drive the motor. The current is the input to this system, but how do we know how to control the current to achieve a certain motor position? This is where feedback control comes in, helping us to **close the loop** between the system's output and the required input, as illustrated in *Figure 5.1*:

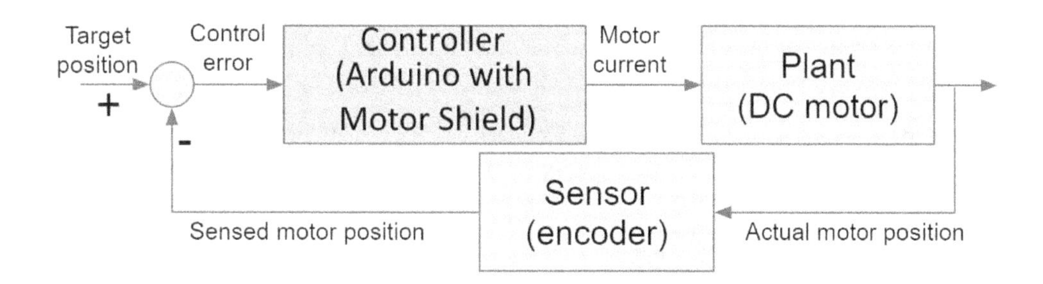


Figure 5.1 – Block diagram of a position control system

In a **feedback control system**, the goal is to drive the output of the **plant** that is being controlled—such as the position of a servo motor—as close as possible to the desired target value. To achieve this, the system continuously monitors the plant's output with a sensor and compares it to the target value. The controller uses the difference between the target and the actual value to compute the required input to the plant, which brings its output closer to the target. You can see how the output is *fed back* to the controller, which is why we call this scheme *feedback control*. In the diagram shown in *Figure 5.1*, you can see how this system forms a closed loop, which is why we often use the term *control loop* to describe a feedback control system.

When we implement a system such as this on an Arduino, we write a program that continuously runs through the three steps of this loop:

- Measure the actual value
- Compare the actual and the target value
- Compute the controller output and send it to the plant

The frequency at which we do this—the **control loop frequency**—is an important parameter. Generally speaking, the higher this frequency, the better the performance of the control system will be. To achieve a high control frequency, it is important to implement efficient and predictable code using the methods we described earlier in this section and to be aware of potential bottlenecks in the form of slowly updating sensors, slow communication, or blocking functions.

## Examples of feedback control systems

Feedback control systems are an important part of almost all physical, technical, and biological systems. They are all around us. Let us quickly look at a few examples.

In a car with cruise control, cruise control is a feedback control system. The car is the plant, its input is motor torque, and its output is velocity. The cruise control system computes the torque that is required to reach and maintain the target velocity. Without cruise control, the human driver acts as the feedback controller.

In an oven, a feedback control system is used to maintain the desired temperature. The oven is the plant, the input is the current to the heating elements, and the output is the internal temperature.

In a battery charger—for example, for a phone—a feedback control system is used to control the charge current. The battery is the plant, the input is the applied voltage, and the output is the charge current. The controller controls the voltage to keep the charge current at the desired level.

These are just a few examples, but you can find feedback control systems in almost any physical system. In practice, they are often even cascaded. A common way to implement a control system for an electric servo motor, for example, is to have a cascaded system of three control loops. The innermost loop is in charge of controlling the motor current. The output of this controller is the PWM duty cycle to drive the power transistors. One level up is a velocity controller that controls the speed of the motor. Its output is the desired motor current, which in turn is the input for the current controller. And finally, the outermost control loop is the position controller to control the motor position. The output of this controller is the desired motor velocity, which is the input to the velocity control loop.

## The PID control algorithm

By now, you will have asked yourself how the controller block in *Figure 5.1* works. How do we compute the required input to the plant from the difference between the desired and target output values? Many books have been written about this topic, but for many applications in robotics, the answer can be surprisingly simple. The **proportional-integral-derivative** (**PID**) controller is often a great choice for feedback control systems that you encounter when creating robots.

In short, a PID controller forms a weighted sum of the control error itself, its integrated value over time, and its time derivative. The controller uses this sum as the input to the plant. The weighting terms are called the **controller gains**, commonly denoted *kp*, *ki*, and *kd* (see *Figure 5.2*). While the PID control algorithm is simple, in practice it can be difficult and time-consuming to find the right values for the three gains. This process is called *controller tuning*. Even though there are practical procedures to find suitable controller gains quickly, it can often be useful to set some of the gains to `0`. A very simple practical controller—and often a good starting point—is a pure *P* controller (with *ki* and *kd* both zero). Similarly, *PD* and *PI* controllers can be good choices that are still easier to tune than a complete PID controller:

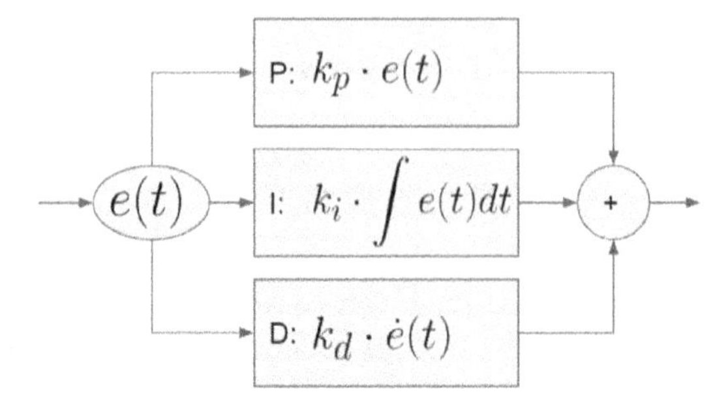


Figure 5.2 – Schematic of the PID control algorithm. The dot above the letter e in the D box denotes the time derivative

Despite its simplicity, there is a lot to learn about PID control in order to apply it successfully in a wide variety of robotics projects. An important thing to note is that it can be difficult in practice to obtain a good (low-noise) *derivative* of the error, and some amount of **low-pass filtering** might be necessary to get used out of the *D* term. Additionally, it is good practice to cap the absolute value of the integral term at some practical limit, since it can otherwise grow unbounded and cause unexpected behavior.

Instead of going deeper into the theoretical aspects of PID control, the next section will show you how to build a DC servo motor that you can use for hands-on experiments with PID control. You can use it to experiment with different gains, control frequencies, feedback delays, and even feedback signal filters.

## Building a DC servo motor

A position-controlled servo motor is a common subsystem of a mobile robot. It can be used to command the joint of a robot arm to a certain position or to accurately drive a wheel to reach a specified position. It is also a great and interactive example of a feedback control system, so let us learn how to build one from the components shown in *Figure 5.3*!

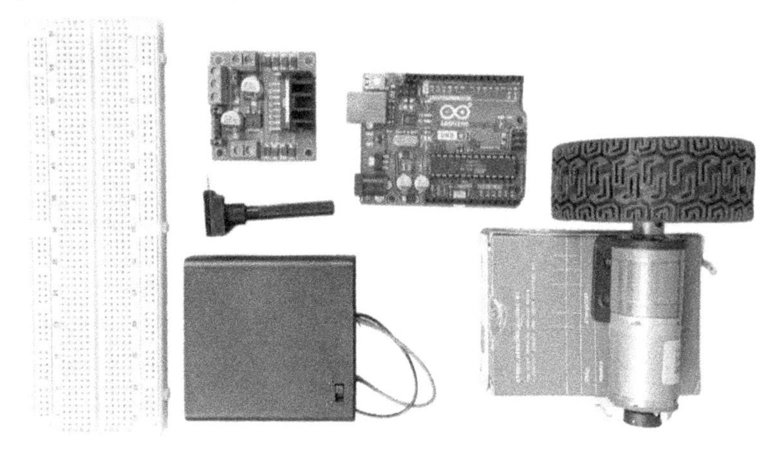

Figure 5.3 – The components needed to implement your own DC servo motor

We will use a battery pack, an L298-based motor driver board, a potentiometer, an Arduino Uno, a geared DC motor with a quadrature encoder, and a wheel to make it easy to turn the motor by hand.

### *Sensing the motor position*

Sensing the system output is crucial for a closed-loop control system. In the case of the DC servo motor, this means sensing the position of the motor shaft. In this example, we use a motor with an integrated incremental quadrature encoder. A quadrature encoder has two digital output signals, commonly referred to as *A* and *B*. Whenever one of them changes from high to low or low to high, we know that the motor has moved one step, and we know in which direction it moved from the state of the other pin (high or low). The diagram in *Figure 5.4* of the two output signals of a quadrature encoder when the motor spins:

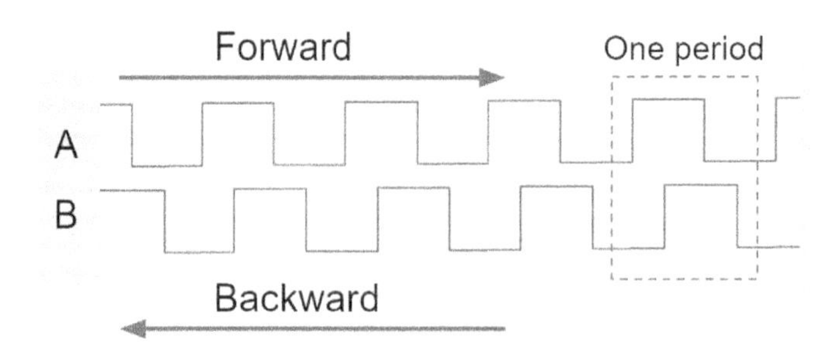


Figure 5.4 – Output signals of a quadrature encoder

When the motor spins forward, the two signals appear from left to right. There are four events in a signal period that all indicate a forward step. If we sense any of these four, we know that the motor moved one step forward:

- A rises, B is low
- B rises, A is high
- A falls, B is high
- B falls, A is low

When the motor spins backward, the signals appear from right to left. This leads to four events per signal period that indicate a backward step. If we sense any of these four, we know that the motor moved one step backward:

- B rises, A is low
- A rises, B is high
- B falls, A is high
- A falls, B is low

Interrupts are the best way to sense changes in signals (called rising and falling edges) with an Arduino microcontroller. Since these rising and falling edge events can happen at high frequencies when the motor spins fast, we need to be careful to implement **interrupt service routines** (**ISRs**) efficiently. There are many ways to implement them, but the following is an example of an implementation that is both very efficient and quite readable. Let's break it up into a few smaller chunks. As usual, we will define a few useful global variables first:

```
// Use Arduino Uno interrupt pins.
// Pins 2 and 3 are our only options.
const int pin_A = 2;
const int pin_B = 3;
```

```
// Variables that are changed inside ISRs should always
// be volatile.
// Variables input pin states.
volatile bool state_a = LOW;
volatile bool state_b = LOW;
// Encoder position.
volatile long encoder_position = 0;
```

To avoid tricky problems that can arise from unwanted compiler optimization or even data corruption, we make sure to qualify variables that are used by ISRs as `volatile`.

In the `setup()` function, we attach an ISR to the rising edges of both interrupt pins. These pins are connected to the two outputs of the quadrature encoder—A and B—respectively:

```
void setup() {
  Serial.begin(115200);
  attachInterrupt(digitalPinToInterrupt(pin_A), risingIsrA,
RISING);
  attachInterrupt(digitalPinToInterrupt(pin_B), risingIsrB,
RISING);
}
```

Since reading the encoder signals is entirely interrupt-driven, there is nothing we need to do inside the `loop()` function. All we do here is print out the current value of the `encoder_position` variable every 50 ms. The fact that we use a `delay()` function in the `loop()` function might look alarming, but is not a problem here. The ISRs will execute just fine even if the main loop is currently blocked by a `delay()` function:

```
void loop() {
  Serial.print("Position:");
  Serial.println(encoder_position);
  // Blocking call, but that's OK.
  delay(50);
}
```

In the ISRs, we update the variable that represents the state of their respective pin, and then we attach the ISR that handles the other edge of the pin. Lastly, we either increment or decrement the encoder position, depending on which of the eight states that we had previously listed we find ourselves in. This is efficiently implemented as a `switch()` statement, like the ones we saw when we implemented state machines. Since both pins A and B can fall or rise, we need to implement four ISRs. The ones for pin A are shown in the following code block, and the ones for pin B look very similar. You can find the complete code in this chapter's folder on GitHub:

```
void risingIsrA() {
  state_a = HIGH;
  attachInterrupt(digitalPinToInterrupt(pin_A), fallingIsrA,
FALLING);
  switch (state_b) {
    case LOW:
     // Encoder moved forward.
      encoder_position++;
      break;
    case HIGH:
      // Encoder moved backwards.
      encoder_position--;
      break;
  }
}


void fallingIsrA() {
  state_a = LOW;
  attachInterrupt(digitalPinToInterrupt(pin_A), risingIsrA,
RISING);
  switch (state_b) {
    case LOW:
      // Encoder moved backwards.
      encoder_position--;
      break;
    case HIGH:
      // Encoder moved forward.
      encoder_position++;
      break;
  }
}
```

With this code loaded and the encoder connected to your Arduino, you can open the Serial Monitor and spin the motor. You will see the position value increase or decrease, depending on the direction you turn it. You can also use the Serial Plotter to see a running live plot of the motor position, as shown in *Figure 5.5*:

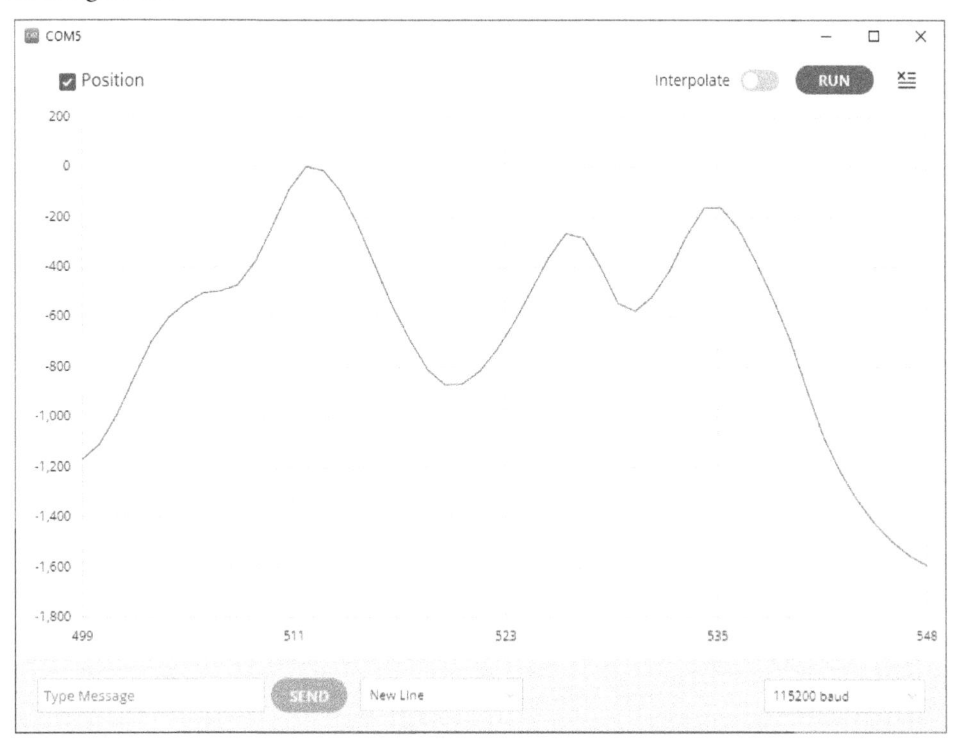


Figure 5.5 – Example output of the preceding code in the Serial Plotter of the Arduino IDE when the motor is spun by hand

You now know how to use the quadrature encoder of a motor. To make a servo motor, however, we also need to drive the motor!

## *Driving a DC motor without the Arduino motor shield*

We already know how to drive a DC motor with the Arduino motor shield. However, for this setup, we will not use the Arduino motor shield for two reasons:

- The motor shield occupies pin 3 for PWM A, but we need this pin as an interrupt pin to read the quadrature encoder.
- It uses pins 3 and 11 for PWM A and PWM B. On these two pins, the PWM signal that the Arduino produces has a frequency of 490 Hz. The PWM frequency on PWM pins 5 and 6 is twice that (980 Hz), which is much better for DC motor control.

There are many alternatives to the motor shield that use the same driver chip but give us more flexibility to select which pins we want to use such as this one, for example: `https://bc-robotics.com/shop/l298n-motor-driver-board/`. Boards such as this have a slightly different interface: instead of one direction input per motor, they have two. The voltage level on the first input (high or low) determines the motor direction. If the second direction pin is driven to the opposite level (low or high, respectively), the motor spins when we enable it via the ENABLE input. If both pins are driven to the same level, the motor brakes when enabled. By sending a PWM signal to the ENABLE input, we can control the motor speed, just as with the Arduino motor shield.

To control the motor via this board, we add the following global variables and the `setMotorPwm()` function to our program:

```
const int dir_pin_0 = 8;
const int dir_pin_1 = 9;
const int pwm_pin = 5;

void setMotorPwm(int motor_pwm) {
  // Control motor direction pins based on the sign
  // of the motor_pwm parameter.
  if (motor_pwm > 0) {
    // Forward.
    digitalWrite(dir_pin_0, HIGH);
    digitalWrite(dir_pin_1, LOW);
  } else {
    // Backwards.
    digitalWrite(dir_pin_0, LOW);
    digitalWrite(dir_pin_1, HIGH);
    // Invert motor_pwm to make it positive.
    motor_pwm = -motor_pwm;
  }
  // Constrain motor_pwm between 0 and 255 to make it
  // compatible with analogWrite().
  motor_pwm = constrain(motor_pwm, 0, 255);
  analogWrite(pwm_pin, motor_pwm);
}
```

This function takes a desired motor PWM value as a parameter. It sets the motor direction based on the sign of its value. It then uses the absolute value of the `motor_pwm` parameter, constrained between `0` and `255`, as input to the `analogWrite()` function to create the PWM signal that drives the motor.

We also need to make sure to set the two direction pins as outputs in the `setup()` function by adding the following lines to it:

```
pinMode(dir_pin_0, OUTPUT);
pinMode(dir_pin_1, OUTPUT);
```

This is all we need to be able to drive a DC motor with a generic motor driver that lets us select the pins more freely. You can feel free to use a different motor driver and change your version of the `setMotorPwm()` function accordingly.

### Implementing a PID position controller

Now that we know how to read the motor position and how to drive the motor in a way that does not interfere with the interrupt pins, the last missing building block for our DC servo motor is the PID position controller. Let us first add the following global variables to our program:

```
// Controller input.
long target_position = 0;
// Controller output.
float controller_output = 0;
// Output of the i control term.
float i_term = 0;
// Last error for computing the I and D term.
int last_error = 0;
// Variables for timekeeping.
unsigned long last_control_time = 0;
const int control_interval = 10;


// Pins for motor control.
const int dir_pin_0 = 8;
const int dir_pin_1 = 9;
const int pwm_pin = 5;
```

We declare the variables that hold the target position and the controller output *global* to make it easy for another task—the print task—to periodically print them to the Serial port. The `i_term` variable will be used to accumulate the scaled integral of the error over time, and `last_error` will be used in both the calculation of the integral and the derivative of the error term inside the controller. And lastly, we add the familiar timing parameters for the control task. A control interval of 10 ms means that the controller will run at a frequency of 100 Hz.

With these global variables available, we can write a PID controller function as follows:

```
void motorPidController() {
  // Controller gains.
  float kp = 2;
  float ki = 0.0;
  float kd = 150;
  // Compute the error (in encoder counts).
  int error = target_position - encoder_position;
  // Compute the P term.
  float p_term = kp * error;
  // Compute and constrain I term.
  i_term += ki * (error + last_error) / 2 * control_interval;
  i_term = constrain(i_term, -255, 255);
  // Compute D term.
  float d_term = kd * (error - last_error) / control_interval;
  // Compute final output.
  controller_output = p_term + i_term + d_term;
  // Update last_error for the next iteration.
  last_error = error;
  setMotorPwm(controller_output);
}
```

As you would expect, the controller computes the proportional term by scaling the error with `kp`, updates the integral term by scaling the error integral with `ki`, and computes the derivative term by scaling the difference between the current and the last error with `kd`. To avoid the integral term from growing indefinitely, we constrain it between the smallest and the largest feasible outputs of the controller.

The output of the controller is simply the sum of these three terms. Lastly, we update the `last_error` variable so that we have it available in the next iteration and then use the computed controller output as a motor PWM command.

Using the familiar mechanism of cooperative multitasking, we can call this controller periodically from inside the `loop()` function with the following code:

```
void loop() {
  if(millis() - last_control_time >= control_interval) {
    last_control_time += control_interval;
    motorPidController();
```

```
  }
}
```

With all of the hard work done, let us start experimenting with our new PID control test bed!

### *Experimenting with the PID controller*

In order to see what is happening when we run this code, let us also add a print task. We will add the usual global timekeeping variables like so:

```
unsigned long last_print_time = 0;
const int print_interval = 20;
```

And then, we can add the print task inside the `loop()` function, right below the controller task block:

```
  if (millis() - last_print_time >= print_interval) {
    last_print_time += print_interval;
    // Print controller input, output, and position.
    Serial.print("Target_position:");
    Serial.print(target_position);
    Serial.print('\t');
    Serial.print("Actual_position:");
    Serial.print(encoder_position);
    Serial.print('\t');
    Serial.print("Controller_output:");
    Serial.println(controller_output);
  }
```

This will create three lines in the Arduino Serial Plotter live plot that lets you see how well the actual position is tracking the target position and how much the controller is pushing the motor.

If you load this code onto your Arduino, connect the motor, and power it from a sufficiently strong battery, nothing should happen, and you should three flat lines at `0` in the Serial Plotter. Since on power-up the encoder value is `0` and the target value is `0`, the controller output will also be `0`. But if you try to spin the motor by hand, the controller should start fighting you, trying to keep the motor at the 0 position. If instead the motor just starts spinning out of control, the controller pushes in the wrong direction. An easy way to fix this is by inverting the direction of the encoder, which is as easy as swapping the wires for the *A* and *B* signals.

Let us make one more addition that lets us change the target position with a potentiometer. That is as simple as adding the following line inside the controller block of the `loop()` function:

```
target_position = analogRead(A0);
```

Connect a potentiometer to *5V*, *GND*, and *A0*, you have a great setup to get familiar with feedback control systems. Let us do some basic experiments right away. Do not worry if the plots look different with your setup—there are many parameters such as the exact motor, the encoder resolution, and the battery voltage that can affect the system's behavior, but you should be able to reproduce all these experiments at least qualitatively.

For the first experiment, we will set `kp` to a small value and leave `ki` and `kd` at zero:

```
float kp = 0.1;
float ki = 0.0;
float kd = 0.0;
```

When you look at the live plot and turn this potentiometer to change the target position abruptly, you will see that the actual position does follow the target position but is not tracking it very well. It lags behind and never quite reaches the target. This is very typical for a proportional controller with a small gain:

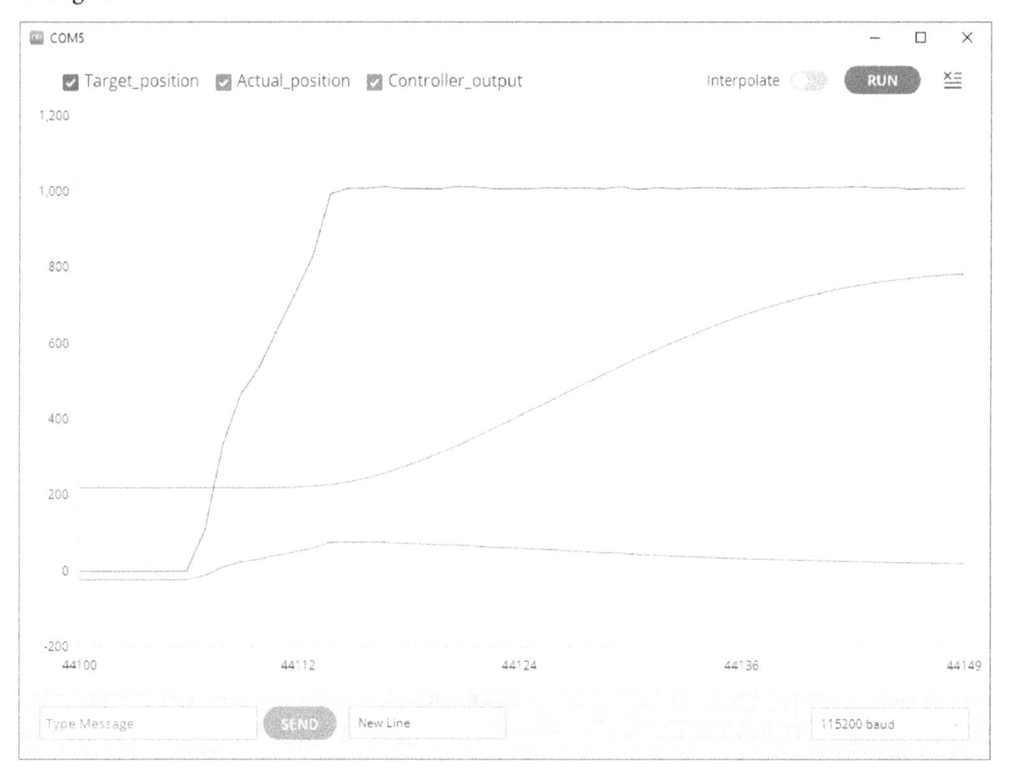


Figure 5.6 – Target position in blue, actual position in red, and the controller output in green for a low-gain P position controller (colored version available on GitHub)

To improve the tracking performance, let us increase the value of kp by a factor of 10:

```
float kp = 1.0;
float ki = 0.0;
float kd = 0.0;
```

You will see a better tracking performance, but you might also see some overshoot and oscillations. This is normal for a high-gain P controller, but overshoot and oscillations are almost never desired. You can also see that the controller output is much larger to create a stronger reaction of the motor:

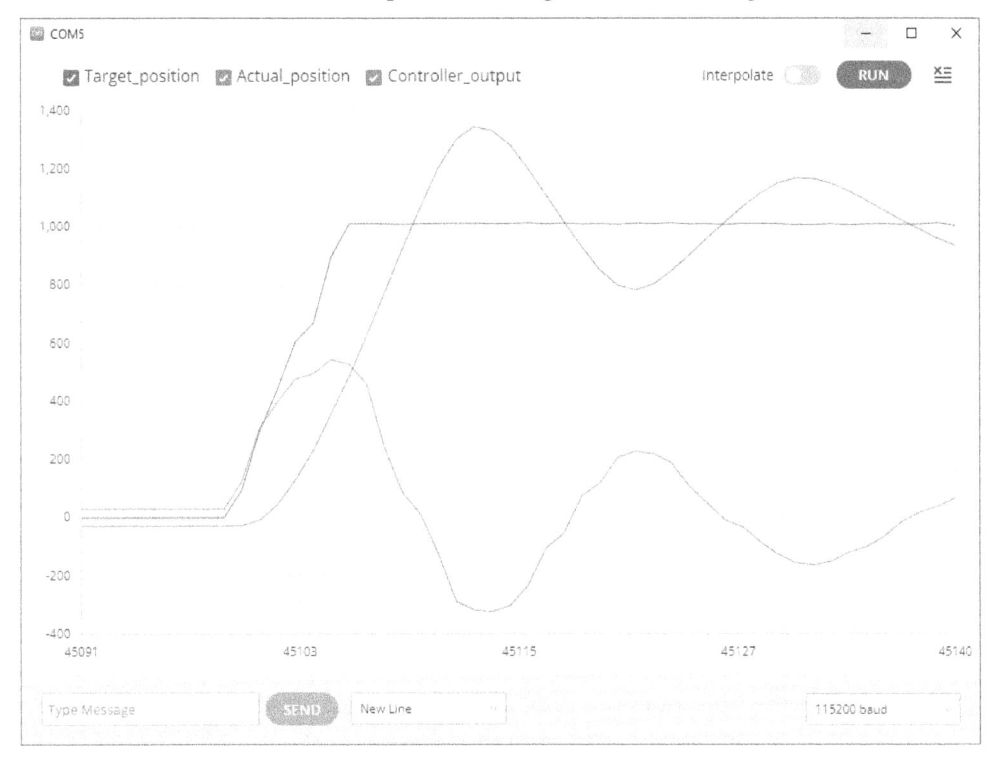


Figure 5.7 – The same experiment with a higher-gain P position controller (colored version available on GitHub)

We can use the *D* part of the PID controller to decrease the amount of overshoot and get rid of the oscillations:

```
float kp = 1.0;
float ki = 0.0;
float kd = 60.0;
```

If we increase `kd`, tracking performance is still good, but overshoot and oscillation are both significantly reduced:

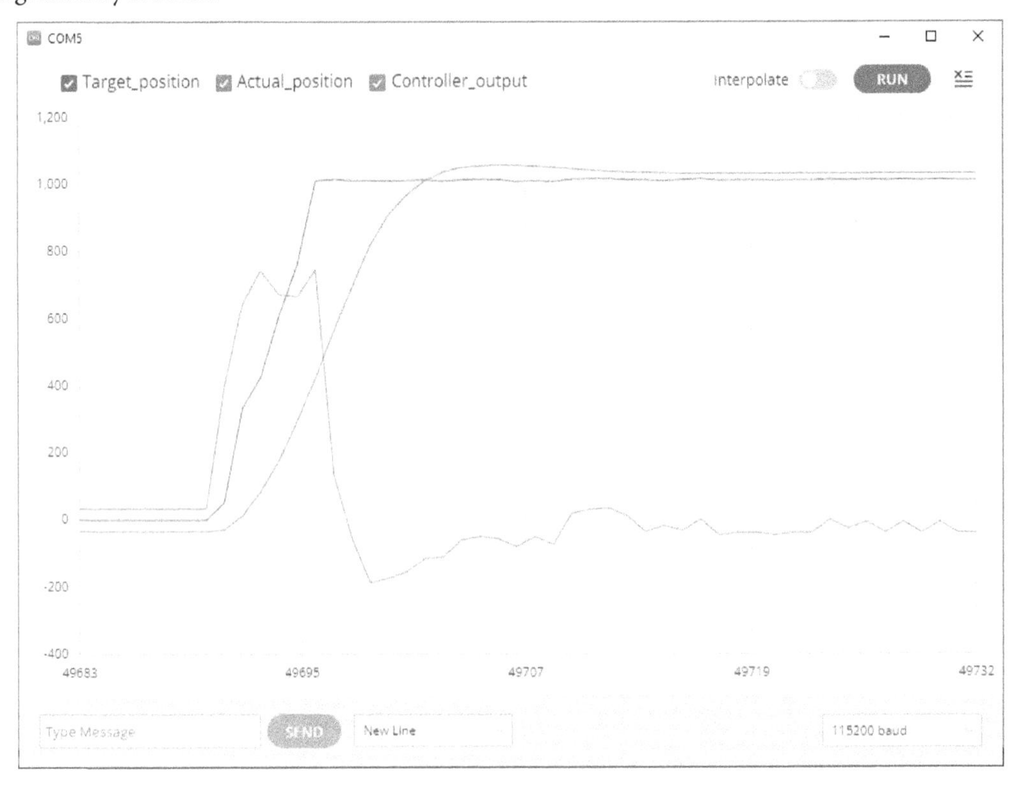


Figure 5.8 – With a well-tuned PD controller, the actual position in red is tracking the target position in blue closely, with a small overshoot and no oscillation (colored version available on GitHub)

You might have noticed that the controller output looks a lot noisier than with the pure P controller. This is caused by how we calculate the error's derivative from a low-resolution position encoder. This process produces a noisy estimate of the motor speed at best, and thus causes a noisy controller output. It is not a problem in our little example, but it can become one in real-life applications. If an encoder with higher resolution is not available, a common mitigation to the noise problem is to apply a low-pass filter to the velocity estimate. This inherently introduces signal lag, which can be quite problematic. Another workaround is to use a sensor that directly *measures* the velocity of the controlled plant instead of using noisy numerical differentiation. However, the required sensors can be large and pricy.

The purpose of the integral term is to compensate for a **steady-state error**. If, for example, there were a spring pulling your motor always in one direction, a PD controller could never achieve the desired target position. An integral controller term could, which is why they are typically used in systems that exhibit a noticeable steady state error with only a PD controller. Our motor control setup has nothing that pulls it in one direction or the other, so increasing `ki` from zero would not actually bring us any benefits. Feel free to try it out, though!

There are many experiments you can use this setup for to strengthen your understanding and build your intuition for control systems:

- What happens when you set any of the three controller gains too high?
- What is the effect of the control loop frequency (the value of the `control_interval` parameter)?
- How does the system behavior change when you use a different voltage to drive the motor?
- Generate the target position automatically (for example, as a time-based square or sine wave) instead of using the potentiometer. This allows a more direct comparison of controller performance across parameter changes.
- Plot the proportional, derivative, and integral terms separately to see which contributes the most to the controller output in each situation.

Building a practical understanding and good intuition for feedback control systems are extremely useful skills to have when it comes to building robots. It is worth spending some time on these and any other experiments you can think of to learn as much as you can about applied control theory.

# Summary

In this chapter, we learned many important aspects of robot programming with Arduino. Cooperative multitasking and a good understanding of state machines are important for almost any reboot program that needs to run on a microcontroller. A good grasp and intuition for feedback control systems are also very useful when building robots. To get you there, we built a PID position-controlled DC servo motor that allows you to experiment with parameters, code, and hardware changes to solidify your understanding of feedback control systems.

In the next chapter, we will focus on object orientation, a very powerful feature of the C++ language that allows us to reuse much of our code and that simplifies using code written by others. You will learn how to write your own **object-oriented** (**OO**) Arduino libraries that can be used by anyone.

Instead of the *Further reading* section on this topic, it is worth spending some time experimenting with your DC servo motor. You can test the effects of different parameters, control frequencies, and battery voltages. You will discover many interesting effects that can guide you in your further studies of this topic.

# 6

# Understanding Object-Oriented Programming and Creating Arduino Libraries

**Object-Oriented Programming** (**OOP**) is one of the excellent features of the C++ language. It is widely used throughout the Arduino ecosystem, particularly when it comes to reusable Arduino **libraries**. However, it is also an advanced feature that can be a little confusing at first. This chapter will introduce you to the concept behind OOP, guide you through writing your first object-oriented program, and show you how to create an Arduino library from scratch, which you can then use in your own projects or even share with the wider Arduino community.

This chapter is structured into the following main sections:

- Understanding the OOP paradigm
- Writing your first class
- Turning your class into an Arduino library
- Working with third-party Arduino libraries

## Technical requirements

This is a software-heavy chapter, and all you need to follow along and try the examples yourself is your PC, your Arduino Uno, and a USB cable to connect the two. For one optional example application, you will additionally need a solderless breadboard, jumper wires, one or more LEDs, and a 220-ohm resistor for each LED.

## The OOP paradigm

OOP is a powerful concept that solves several problems that you might come across when you develop Arduino programs. The most obvious one is the ability to reuse code that you have already written, without the need to copy and paste it throughout your program. Reusing code without having to duplicate it makes it much easier to maintain your code in the future. For example, if you want to make a slight change to an implementation, you only need to make the change in one place instead of having to make it in every program that uses that piece of code.

Imagine, for example, that you have a robot with dozens of LEDs that need to blink at different frequencies. Instead of implementing the same blink logic for each of the LEDs and creating a lot of redundant code, OOP lets you implement the logic only once and then reuse the same code for each LED. If you want to make a change to the way the blinking works, you only need to make it in one place, and it is automatically applied to all the LEDs. There are different flavors of OOP, and we will use what is called *class-based OOP* for our Arduino programs.

In class-based OOP, there are two main concepts that we need to understand: **classes** and **objects**.

In the LED example, all the variables and functions that are needed to blink an LED will be implemented in a piece of code that is called a class. Let us call this class the `Blinker` class. You can think of this class as a detailed how-to that describes to your Arduino, step by step, what it needs to do to blink an LED. This is similar to a function, but a class is a more general how-to that can contain many functions and variables.

We can make use of all this knowledge on how to blink an LED by creating an *instance* of the `Blinker` class. An instance of a class is also called an *object*. Let us call this object `red_blinker`. Using the `red_blinker` object to blink a red LED will require very little code because we can tell our Arduino that all it needs to know about how to use `red_blinker` is already described in the `Blinker` class. The powerful part of this approach is that we can instantiate many more blinker objects, for example, `green_blinker`, `yellow_blinker`, and `blue_blinker`, and they can all be used to control different LEDs connected to different pins, blinking at different frequencies, with minimal additional code.

Another extremely useful feature of OOP is **encapsulation**. Everything that the program needs to know about blinking an LED is encapsulated inside the `Blinker` class, and as programmers, we have full control over how the rest of the program can interact with a blinker object. This can be especially useful for preventing programming errors because the compiler will notice if our program tries to use an object in an unintended way and will generate a compile-time error.

With this very brief introduction out of the way, let us put all of this into practice and implement a basic blinker class.

## Writing and using the Blinker class

In the previous chapter, we learned how to write code that can independently blink an LED and print a message using the cooperative multitasking framework. In this section, we will create the object-oriented version of this code, but only for the blinking part, to keep things a little more concise. At first sight, the object-oriented version might not look like a better way of doing it since it requires even more code. However, if you think about how to scale this code up to blinking many LEDs at different frequencies, the object-oriented code will clearly become the better choice. For reference, the following is the blink part of the state machine example from the last chapter:

```
unsigned long last_blink_time;
int blink_interval = 200;

void setup() {
  Serial.begin(115200);
  pinMode(LED_BUILTIN, OUTPUT);
  last_blink_time = millis();
}

void loop() {
  if (millis() - last_blink_time >= blink_interval) {
    last_blink_time += blink_interval;
    digitalWrite(LED_BUILTIN, !digitalRead(LED_BUILTIN));
  }
```

To turn this into an object-oriented version, we can transfer all the variables and code that are only relevant to blinking the LED into a `Blinker` class that we will create at the top of the program. This will create the desired encapsulation and will allow us to instantiate as many blinkers as we like without having to add new functions and global variables for timekeeping. Take a look at the following code before we go through it in much more detail:

```
// Begin of class definition.
class Blinker {
  // Accessible outside the class.
public:
  Blinker(int led_pin, int blink_interval) {
    // Use constructor arguments to set private variables.
    led_pin_ = led_pin;
    blink_interval_ = blink_interval;
```

```
    last_blink_time_ = millis();
  }

  // Initialize required hardware.
  void begin() {
    pinMode(led_pin_, OUTPUT);
  }


  void blink() {
    if (millis() - last_blink_time_ >= blink_interval_) {
      last_blink_time_ += blink_interval_;
      blink_task();
    }
  }


  // A public set method to change a private variable.
  void set_blink_interval(int blink_interval) {
    blink_interval_ = blink_interval;
  }


  // Only accessible within the class itself.
private:
  unsigned long last_blink_time_;
  int blink_interval_;
  int led_pin_;


  void blink_task() {
    digitalWrite(led_pin_, !digitalRead(led_pin_));
  }
};
// End of class definition.
```

After the class definition comes the rest of the program, which is very short now:

```
// Instantiate a Blinker object called myBlinker.
Blinker myBlinker(13, 200);

void setup() {
  myBlinker.begin();
}

void loop() {
  myBlinker.blink();
}
```

Now let us walk through the entire program step by step!

At the very beginning of the program, we start the definition of the class with the `class` keyword, the name that we want to give this class, and open curly braces. At the very end of the class definition, we close the curly braces and add a semicolon. Everything inside this pair of curly braces, the class definition, is divided into two parts: the **public** and the **private** elements of the class definition. The public block begins with the `public` keyword and the private block with the `private` keyword.

## Public and private class members

The variables and functions inside a class definition are called its **members**. They can be either public or private, depending on the part of the class definition they are defined in. If we instantiate an object based on the class definition, this object will have exactly these public and private members. Public members of an object can be used or called by the main program. To call a public member function or access a public member variable of an object, you simply type the name of the object, followed by a dot (period), and the name of the function or variable. In the `loop()` function of our example, we can call the `blink()` function of the `myBlinker` object with the line `myBlinker.blink()` because `blink()` is a public member function. In contrast, private members can only be used or called inside the class definition itself; they are not visible or accessible outside of it—they are private to the class. This means you could not add this additional line in the `loop()` function:

```
myBlinker.blink_task();
```

The compiler would present you with an error message to inform you that you are trying to use a private member function outside the class definition, and the compilation process would fail.

**Compilation error: 'void Blinker::blink_task()' is private within this context**

The public functions of a class are often called its **Application Programming Interface** (**API**), or simply **Interface** because they are how the application program can interface with an object of that class. The private functions represent the inner workings of the class that are hidden from the application code outside the class, in the spirit of encapsulation. There is no strict rule as to what type of function should be public and what should be private, but it is typically good practice to keep the public interface to a minimum and make as many functions and variables as possible inside a class private. This helps to prevent unintended use of your class's member functions that could lead to runtime errors. Instead, the compiler will cause compile-time errors that are much quicker to find and eliminate. Keeping the API concise further allows you to later change the inner workings of your class (that is, the private member functions) without breaking any parts of other code that uses your class since that code only depends on the (small) public API.

Every class needs at least one public function that we can use in our program to instantiate a new object of this class. This is a special function called the **constructor** because it quite literally constructs a new object. The constructor has the same name as the class, and it can take in arguments like any other function. In our example, the constructor of the `Blinker` class takes in the pin that the LED is attached to and the interval in milliseconds between switching the LED on or off. It is entirely your design choice what arguments go into the constructor. It is generally a good idea to have the constructor initialize all the variables that need to be set only once and use dedicated public setter functions, called **setters** (or more formally, **mutators**), for private member variables that might need to be changed throughout the runtime of the program. Public functions to read private member variables are called **getters** (or more formally, **accessors**). The constructor can initialize private variables either to hard-coded default values or to the values passed to it. If your class does not specify a constructor, a default constructor will automatically be generated at compile time.

Oftentimes it is intuitive to give arguments that you pass to the constructor the same name as the private variable that you want to initialize. For example, if we want to pass in an argument to initialize a private variable called `led_pin`, that parameter could also be called `led_pin`. To avoid too much confusion that can be caused by using the same name for variable and argument, it is good practice and a widely accepted convention to end the names of private member variables with an underscore. That way, a line such as `led_pin_ = led_pin;` in the constructor is easy to read and makes intuitive sense. You can read it as *Set the private member variable led_pin_ to the value of the constructor argument led_pin.*

Our `Blinker` class has three more public member functions that make up its interface or API. The first one is the `begin()` function. For Arduino classes, we try to put all the code that initializes hardware inside the class's `begin()` function. This function is called only once for each object, inside the `setup` function. The second API function is `blink()`. It needs to be called periodically in the program's `loop()` function, checks if it is time to change the LED state, and if yes, it runs the blink state machine that we already know. The last one is the `set_blink_interval()` function that our program can use to change the value of a `Blinker` object's private variable, `blink_interval_`.

You might wonder, what is the benefit of using a dedicated public function to change a private variable, as opposed to simply making the variable public instead? In this very simple example, it is indeed hard to justify, but there are in general several good reasons for this pattern. For example, if you want to check that only valid values can be set, you can add a validity check to the setter function. Or, if you want to keep track of how often the value of the variable was changed, you can add a counter to the setter function.

These four functions make up the API of your `Blinker` class, and that is all that the main program needs to know about to use all the functionality of this class. All the other variables and code that we needed in the non-OOP implementation of the blinker are private members of the `Blinker` class. The only change we made is to end all private variable names with an underscore.

## Using our class definition

With the definition of our brand-new `Blinker` class on hand, let us look at how to use this class in our Arduino program. Right below the end of the class definition, in the space where we usually define global variables, we can now instantiate a global Blinker object. To do this, we first type the name of the class that we want to instantiate an object of, then a space, and then the name of the object we want to create followed by the constructor parameters in parentheses:

```
Blinker myBlinker(13, 200);
```

You can read the line as follows: *Create an object of the Blinker class named myBlinker by calling the constructor of the Blinker class with the parameters 13 for the pin number and 200 for the blink interval.* After this line is executed, you have a new object called `myBlinker` that has its `led_pin_` value set to `13` and its `blink_interval_` value set to `200`, and you can use the public interface of this object in your code. In our example, we simply call its `blink()` function inside the `loop()` function to make the LED on pin 13 blink with a 200 ms on- and off-time.

However, since our class also has a public function to control the blink interval, we can make this program a little more interactive by adding the following line to the `loop()` function:

```
myBlinker.set_blink_inerval(analogRead(A0));
```

If you now connect a potentiometer to your Arduino as we have done before, with the outer two pins connected to GND and 5V and the center pin connected to the A0 pin, you will see the blink interval change from 0 ms (LED at half brightness) to 1,023 ms, depending on the position of the potentiometer.

The little `Blinker` class can be very useful to control a simple status LED of your robot, where the blink interval can signal different conditions. For example, if it blinks at an interval of 500 ms, everything is OK. If the robot encounters a certain problem, you can use the `set_blink_interval()` function to reduce the interval and make the LED blink more rapidly to signal the problem, yet another robot state could be signaled by the LED blinking much slower, and so on.

## The power of OOP

Now comes the best part: what if you had another LED on pin 12 that you want to blink in a 350 ms interval? All you need to do is add three lines of code to your program! Under the instantiation of `myBlinker`, simply instantiate another `Blinker` from the same class definition like so:

```
Blinker myOtherBlinker(12, 350);
```

Then, add a call to its `begin()` function inside `setup()` like so:

```
myOtherBlinker.begin();
```

Call its `blink()` function inside the `loop()` function like so:

```
myOtherBlinker.blink();
```

That is all that you need to do! If you now connect a second LED to pin 12 as shown in *Figure 6.1*, you will see both LEDs blink with a 200 ms and a 350 ms period, respectively.

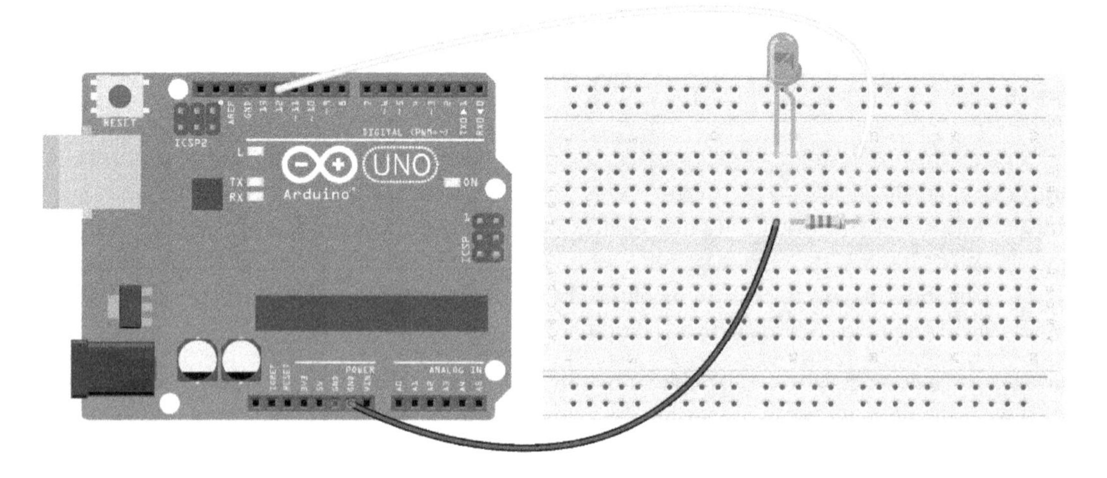


Figure 6.1 – Hardware setup to blink the onboard LED on pin 13 and a second LED on pin 12

Using the same method, you can easily add as many blinkers as you need to control as many LEDs as you want at any blink interval you like.

# Turning your class into an Arduino library

A class is a bundle of variables and functions that encapsulates all the code needed for an object of this class to perform certain tasks. The LED blinking example is quite simple to illustrate the core concepts, but you can easily imagine how classes can get much more complex, with many more member variables, functions, and a richer interface. Many parts of examples in previous chapters could be rewritten as classes for future reuse, for example, the PID position controller and the motor driver code.

In the last section, we saw how we can write a class inside our Arduino program and then use it to instantiate many objects of that class, but we have not seen yet how to reuse the class definition in another program. Being able to reuse code across programs, and even share it between different programmers so they can all use it in their programs, is one of the most useful features of OOP in the Arduino ecosystem. For this reason, the majority of Arduino libraries, both the official ones that come with the IDE and third-party ones that you can find on GitHub, are implemented as classes. So, let us learn how to turn our own `Blinker` class into a proper Arduino library that you can reuse in every Arduino project and share with other Arduino programmers. Once you have your `Blinker` library, you will never have to think about how to make an LED blink again.

To create an Arduino library from the `Blinker` class, we need to move the code from the Arduino program into two new files: the **source file** and the **header file**. You can think of the header file as the table of contents of the class definition. It is used by the compiler to know what it can find in the source file, and the source file is the actual content.

## The header file

First, let us create the header file. To do this, you can use any text editor that you like to create a new file called `Blinker.h`. The content of this file needs to look as follows:

```
#ifndef BLINKER_H_
#define BLINKER_H_

#include "Arduino.h"

class Blinker {
public:
  Blinker(int led_pin, int blink_interval);
  void begin();
  void blink();
  void set_blink_interval(int blink_interval);

private:
  unsigned long last_blink_time_;
  int blink_interval_;
  int led_pin_;
```

```
  void blink_task();
};

#endif  // BLINKER_H_
```

The class definition will look very familiar to you. It looks just like our earlier example, except that all the content of the function definitions was removed. Only the function **signatures**, with a semicolon at the end, remain. This is because the header file is like a table of content of the class; it does not contain the actual implementation. Above the class definition, we added the `#include "Arduino.h"` line to be able to use all the standard definitions of the Arduino language. If we write code inside the Arduino IDE, we can omit this line and it will be added automatically during the compilation process. However, since the header file of our library exists outside the Arduino IDE, we need to add it in order for the compiler to compile it successfully. Lastly, you can see two more lines on top and one line at the very end of the file; these three lines of **preprocessor directives** implement a so-called **include guard**. They are not C++ code; they just make sure that this header file is not accidentally included more than once, which could cause compiler problems that might be hard to debug. Adding these lines is not strictly necessary in this example, but it is good practice.

## The source file

The source file contains the missing implantations of all the functions that are listed in the header file. To create it, open a new file in a text editor of your choice, name it `Blinker.cpp`, and add the following content to it:

```
#include "Blinker.h"

Blinker::Blinker(int led_pin, int blink_interval) {
  led_pin_ = led_pin;
  blink_interval_ = blink_interval;
  last_blink_time_ = millis();
}

void Blinker::begin() {
  pinMode(led_pin_, OUTPUT);
}

void Blinker::blink() {
  if (millis() - last_blink_time_ >= blink_interval_) {
    last_blink_time_ += blink_interval_;
```

```
    blink_task();
  }
}

void Blinker::set_blink_interval(int blink_interval) {
  blink_interval_ = blink_interval;
}

void Blinker::blink_task() {
  digitalWrite(led_pin_, !digitalRead(led_pin_));
}
```

All these functions will look very familiar to you as well as they are exactly the same our previous example where the class was defined within the Arduino program. The only change is that all function names are prefixed with the name of the class that they belong to and two colons. This is needed to let the compiler know that these functions belong to the `Blinker` class that was defined in the header function. Lastly, to tie the header and the source file together, the source file begins with a line that tells the preprocessor to include the `Blinker.h` header file.

## Using the library in the Arduino IDE

To turn the header and the source files into an Arduino library, all that is left to do is to place them in the right location. When you installed the Arduino IDE on your computer, it created a folder called `Arduino`. This is the same folder in which all your Arduino programs (sketches) are stored. The `Arduino` folder has a subfolder called `libraries`. In the `libraries` folder, create a new subfolder called `Blinker` and move both the `Blinker.h` header file and the `Blinker.cpp` source file there. Now close all windows of the Arduino IDE and start the IDE again, and your newly created Blinker library will be available to all your Arduino programs.

Let us put it to the test right away. Create a new sketch that looks like this:

```
#include "Blinker.h"

Blinker myBlinker(13, 50);

void setup() {
  myBlinker.begin();
}

void loop() {
```

```
    myBlinker.blink();
}
```

The blink program has become incredibly simple. Let us recall for a second that behind the unassuming call to `myBlinker.blink()` hides a high-quality, non-blocking implementation that supports cooperative multitasking. We have come a long way from the very first blink program in our first chapter.

After you include the `Blinker.h` header file of your library, you can use all the functions of its interface. If you compile, transfer, and run this sketch on your Arduino Uno, you will see the LED on pin 13 blink very rapidly. Congratulations, you have successfully created and used your first Arduino library! Now you can blink as many LEDs as you want in any of your robot programs with minimal effort.

## Adding examples

A useful feature of Arduino libraries is the ability to add examples that you can refer to later in case you need a refresher on how to use the library. To add a minimal example to your library, create a subfolder called `examples` in it (on the same level as the header and source file). In this folder, create a new subfolder called `minimal_example`, and in this folder a new file called `minimal_example.ino`. You can simply paste the code from the last example into this file.

If you now restart the Arduino IDE, you can easily find this example from the drop-down menu under **File** | **Examples**.

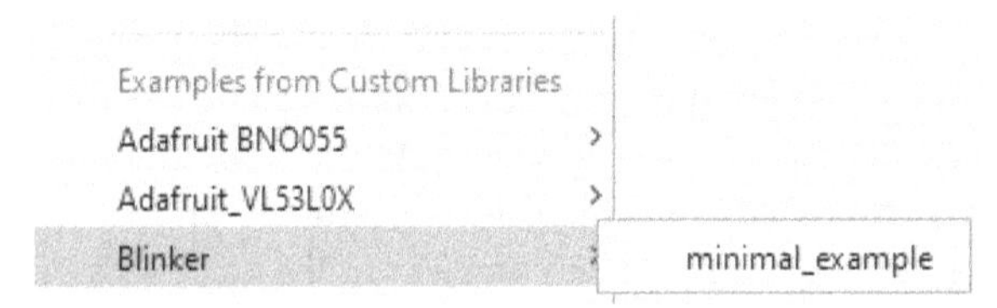


Figure 6.2 – The example program you created in the Examples drop-down menu

You can add multiple examples to your libraries to highlight specific uses. This makes more sense with libraries that are a little more complex than our blinker example, which only has one use case.

Congratulations again, you now know how to create your own Arduino libraries from scratch to reuse your code across projects and share it with other roboticists.

# Working with third-party Arduino libraries

By now you have a good understanding of the purpose and structure of Arduino libraries. They are typically a header file and a source file, and sometimes a folder with examples and a file with meta info about the library that needs to be placed in your `libraries` folder.

## Adding library files manually

There are several ways to use third-party libraries in your project. The most straightforward way is to do the same as what you did with your own library. Once you download all the files of the library, place them in a subfolder with the name of your library within your `libraries` folder, and after the next start of your Arduino IDE, you will be able to use the library and all the examples that come with it.

## Using the IDE's Library Manager

An easier way to install libraries is through the Arduino IDE's built-in Library Manager. All official Arduino libraries and many major third-party libraries are available through this useful tool. If you are looking for a specific library, open the **LIBRARY MANAGER** by clicking **Tools | Manage Libraries...** and use the search bar to find the library you are looking for. You can also use the library icon on the left side of the IDE to open the **LIBRARY MANAGER**. If the library you are looking for is available, simply click the **Install** button.

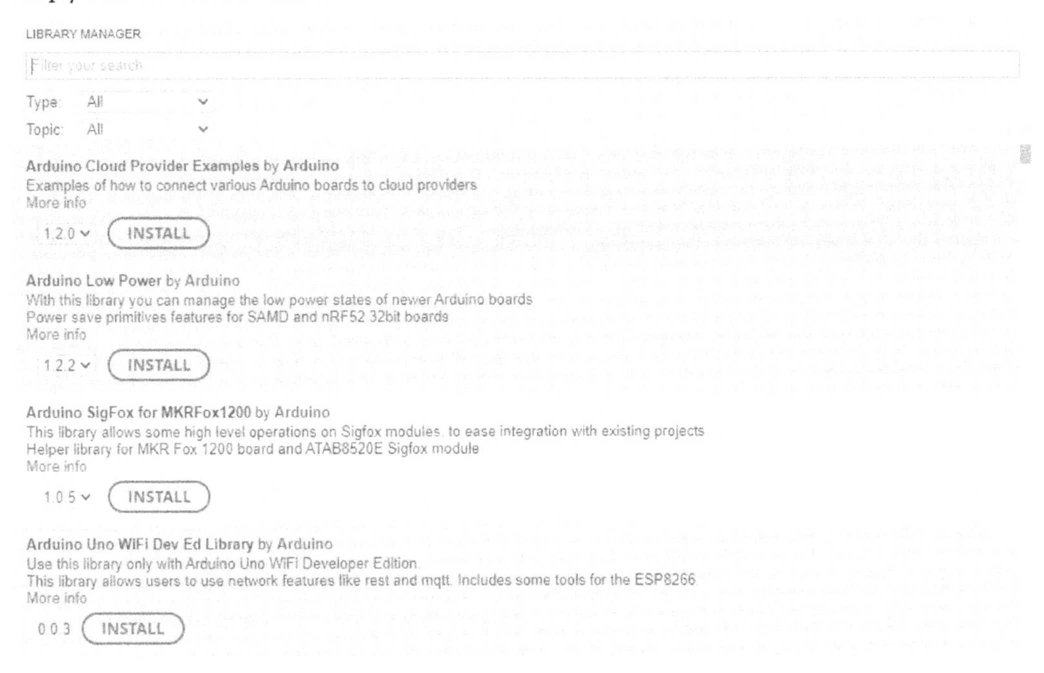


Figure 6.3 – The Arduino IDE's built-in library manager

You can even add your own library to the Arduino Library Manager so that anyone who uses Arduino can easily install it this way. To do this, you need to host your library on GitHub and submit a pull request to the library registry repository to add a link to your library: `https://github.com/arduino/library-registry`. If you have a library that you think is useful for others, this is one of the easiest ways to contribute to the Arduino ecosystem. If you are not familiar with Git and

GitHub, this can be a great exercise and an interesting experience that introduces you to the workflow of open source software development.

# Summary

If this was the first time that you learned about **object-oriented programming** (**OOP**), then you have just learned a major new skill in your journey toward becoming a proficient robot programmer. The ability to divide your program into modular and reusable classes will improve your efficiency as a robot developer as it eliminates many potential sources of error when it comes to reusing code by means of copy and paste. It also makes it much easier and more efficient to maintain your codebase by allowing you to create universal building blocks, such as the Blinker library that we wrote in this chapter. You have learned how to write an object-oriented Arduino library from scratch and how to use it in the Arduino IDE. You also learned two separate ways to use Arduino libraries that were written by others.

In the next chapter, we will learn how to develop our robot programs in a way that makes them easy to test and give us confidence that they work as intended. We will look at techniques for debugging common problems, including the advanced use of the Arduino IDE's built-in Serial Monitor and Serial Plotter.

# Further reading

We touched upon many topics in this chapter. Following are some resources that can help you dive deeper into several of them:

- A good, more detailed tutorial on OOP in C++: `https://www.w3schools.com/cpp/cpp_oop.asp`
- Tips for designing your library from the official Arduino documentation: `https://docs.arduino.cc/learn/contributions/arduino-library-style-guide`
- How to contribute libraries to the Arduino Library Manager: `https://github.com/arduino/library-registry`
- Learn the basics of GitHub to contribute your Arduino Library to the Library Manager: `https://docs.github.com/en/get-started/quickstart/hello-world`
- An in-depth look at C++ classes: `https://en.cppreference.com/w/cpp/language/classes`

# 7

# Testing and Debugging with the Arduino IDE

Testing and debugging are essential parts of every development process. Testing becomes more important and more difficult the more complex a robot is. As we have already seen, even a relatively simple Arduino robot consists of several physical subsystems, such as motors, sensors, batteries, the Arduino board itself, and all the wiring in between. The program that runs on your robot also consists of several components, including hardware interfaces, external libraries, and tasks that need to cooperate with one another to efficiently share CPU resources.

It is common for robot builders, especially less experienced ones, to build an entire robot, write the program for it, and then test the hardware and software all at once. It is very unlikely that everything works as expected on the first attempt, and trying to identify the underlying problems can be very difficult and a major source of frustration in this scenario. This approach is sometimes called **Big Bang Integration** and can lead to a frustrating phenomenon that developers sometimes call **Integration Hell**, where nothing seems to work, and it is unclear why.

A much better approach is to build your robot—both hardware and software—as modular as possible. This allows you to thoroughly test all subsystems individually, making it much easier and more effective to identify and fix problems early on. It also allows you to integrate subsystems one by one, avoiding the problems of Big Bang Integration.

This chapter will introduce general strategies and tools built into the Arduino IDE to efficiently test the subsystems of your robot before putting it all together. The chapter is organized as follows:

- Building modular robot hardware
- Writing testable robot programs
- Using the Arduino IDE for testing and debugging

## Technical requirements

All you need to recreate the examples in this chapter is your Arduino Uno. If you have a logic analyzer on hand, you can use it for this chapter, but it is not required.

## Building modular robot hardware

Modularity is the key to enabling efficient testing of subcomponents to avoid the potentially frustrating experience of Big Bang Integration. This is true for hardware and software, and for the hardware it applies to both the mechanical as well as electrical components of your robot.

### Mechanical components

Problems with the mechanical subsystems of your robot, such as the wheels with their motors or a gripper mechanism, are typically the easiest to troubleshoot. Issues like too much friction, overheating, or a motor turning in the wrong direction are easy to spot without dedicated tools. In addition, the mechanical components of your robot will usually be quite modular in nature without having to put too much thought into them.

Especially during the development phase of your first robot prototype, it is best to keep all mechanisms as simple as possible to make it easy to spot where a particular problem comes from. Ideally, you have a dedicated motor and gearbox for each driven wheel—for example, a dedicated motor for each joint of a robotic arm, a simple gripper mechanism, and no coupling between separate mechanisms. This will make it easy to pinpoint the root cause of problems with your mechanisms and allow you to iterate easily through design modifications. You might find that the motor that you chose for a certain function is not strong enough or too slow. There might be too much friction in a bushing, and you need to upgrade to a ball bearing. Or, the range of motion of an RC servo motor is not large enough and you need to change the geometry of your mechanism or choose a different servo. In any case, these issues will be easier to identify in a simple and modular mechanical system.

Once you are past the initial prototype stage and you really know the requirements of your application, you can start to optimize the mechanical systems with more complex mechanisms. This can allow you to use fewer motors by combining certain functions to make your robot smaller or lighter, faster, or cheaper.

### Electrical components

The same reasons why we want modularity of the mechanical systems also apply to the electrical system of your robot. Here, it is even more important because problems with the electrical subsystems are usually far less obvious than mechanical issues. We might be able to hear and feel mechanical vibrations, for example, but electrical noise requires a costly and complex *oscilloscope* to detect. Similarly, we can feel excessive friction in a mechanism when we move it by hand, but spotting a resistor that is too large can be very difficult even if we have a *digital multimeter* and access to the resistor in question.

### Wiring

By far, most problems with the electrical subsystems of Arduino robots are caused by wiring issues. Connectors, especially when jumper wires and robot motion are involved, can cause intermittent connections and non-deterministic problems. These can be extremely tedious to debug. It is therefore important to use high-quality jumper wires and high-quality solderless breadboards in the prototyping phase to avoid these issues as much as possible.

Even the wires themselves can start to cause problems and even break when they are repeatedly bent, such as wires that span one or more joints of a robot arm. In this case, you want to route the cables such that their motion is minimized and their bending radius is maximized. Using service loops in your cables or even slip rings to avoid motion inside the cables altogether are proven strategies to achieve this. However, you also want your cables to be as short as possible to avoid signal degradation and attenuation of analog and digital signals, as well as power dissipation in power cables.

It is also a smart idea to use a breadboard as a central power distribution hub for all your robot's components, as opposed to building custom wire harnesses that can often be additional sources of problems. If you use soldered connections, make sure that you are familiar with good soldering practices and check every solder joint. The two most important tests are as follows:

- Is the solder joint nice and shiny? If it is matte, this can be an indication of a bad solder joint.
- Can you give the solder joint a decent tug without it deforming or coming apart? A solder joint should be mechanically sturdy; otherwise, it might be a *cold solder joint* and can cause problems later.

If you use connectors that require crimping, try to avoid crimping yourself and opt for pre-crimped leads. Proper crimping is a sensitive process that requires special tools and a lot of experience, and improper crimping is a common cause of faulty electrical connections.

For any connectors that you use, choose gold-plated ones whenever you can. The gold plating reduces the connection resistance and, even more importantly, virtually eliminates the risk of contact corrosion. Otherwise, contact corrosion can slowly degrade the functionality of the connection, causing electrical faults that seem to come out of nowhere. This is especially relevant for outdoor robots.

### Power

We will soon dedicate an entire chapter to the power system of your robot, but let us quickly discuss the types of electrical problems that the power system can cause.

If your robot is powered by a battery, the battery will run low at some point. When the battery voltage drops, some parts of your robot might continue to work just fine, while others stop working or display unexpected behavior. This is called an **undervoltage condition**. For example, some sensors could start to report erroneous readings caused by too low of a supply voltage, or a motor might turn slower than expected, causing a mechanism that relies on synchronized motion to fail. Because different components can start to misbehave at different voltage levels, it might not be obvious at first that the problems simply come from the low battery voltage. It is a clever idea to have your Arduino monitor the battery voltage (via a *voltage divider*) and signal a low battery via an LED or an audio signal. This can save you time looking for problems when all that is needed is charging up your robot's batteries. In *Figure 7.1*, you can see a simple voltage divider circuit that lets the Arduino monitor the voltage of a 9V battery without the risk of getting damaged. The voltage divider is made from two 10k ohm resistors such that the voltage between the midpoint of these resistors and GND is half the battery voltage. The Arduino can safely read this voltage with its analog input pins, and all you need to do to recover the actual battery voltage is to multiply the Arduino's voltage reading by a factor of two. You can design the voltage divider to map any battery voltage to the range between 0V and 5V (or 0V to 3.3V) to make it safe for your Arduino to measure it.

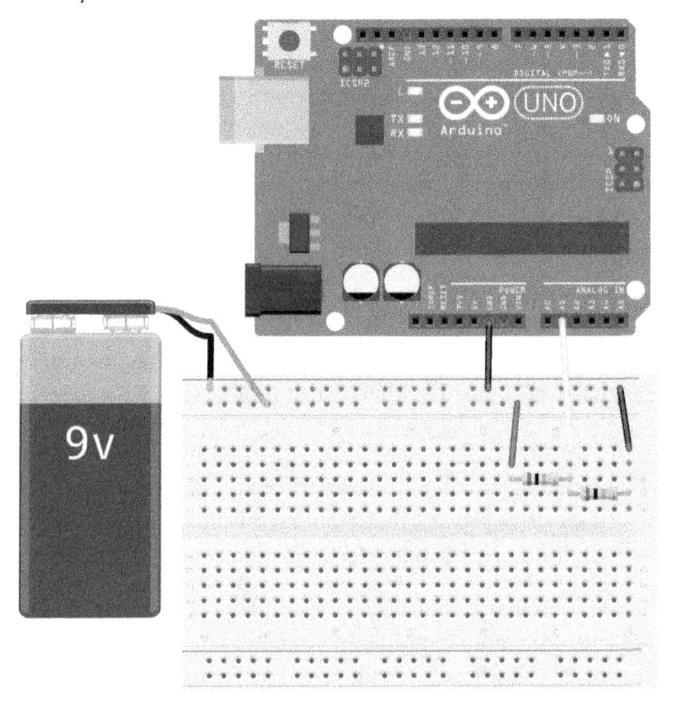


Figure 7.1 – The Arduino monitors the 9V battery's voltage via a 1:2 voltage divider made from two identical resistors in series

Another problem that can arise both when the robot is powered by batteries or an external power supply is a **voltage drop** caused by a large current draw. This happens when the power source is not able to deliver enough power to the robot. A common cause for this is a motor that spins under load or accelerates quickly, drawing a lot of current and asking too much of the power supply, causing its voltage to sag. The symptoms are similar to those of a low battery but occur more sporadically and more abruptly. Sometimes, the voltage can drop so sharply that your Arduino shuts off for a moment.

This situation is called a *brownout*, and the symptoms will be the same as when you press the Arduino's RESET button.

In contrast, the power source of your robot can also supply too much current under certain conditions, damaging components in the current's path that cannot handle it. Most commonly, this happens the very moment the battery is connected to a system that has significant electrical capacitance—for example, in the form of buffer capacitors built into DC/DC converters. Empty capacitors paired with fresh batteries can lead to enormous **inrush currents**, albeit for a very short amount of time. The capacitors that we use in Arduino robots are usually too small to cause problematic inrush currents, but if your system misbehaves the moment you connect the battery, the inrush current might be the problem. There are dedicated (integrated) circuits to limit inrush current. This application note by **Texas Instruments** (**TI**) is a good resource if you want to learn more about this problem: `https://www.ti.com/lit/an/slva670a/slva670a.pdf`.

But not only capacitors can cause power problems; inductors can, too. The current through an inductor (such as the electromagnet inside a relay or the windings inside a motor) cannot stop abruptly. This means if you turn off a relay, for example, the current will continue to flow for a short while. This can cause very high voltage spikes across the now open switch, damaging it and any components on the same circuit. The solution to this problem is the use of **flyback** or **Transient-Voltage-Suppression** (**TVS**) diodes. Motor driver boards always have these components readily integrated to protect against voltage spikes.

### Signals

If something does not work—for example, a digital sensor does not deliver any valid readings—there can be several underlying problems. It could be a wiring problem or a power problem, but it could also be a problem with the signals themselves. In that case, the hardware is perfectly fine, but the communication between the Arduino and the sensor does not work as expected, caused by a software or hardware bug.

If you suspect that there are problems with digital communication, you can double-check all the relevant software parameters. These include, depending on the interface, the baud rate, the bitrate, the SPI mode, or the I2C address. If there are still communication problems, there is an enormously powerful, easy-to-use, and inexpensive tool that can help you find the issue: a *logic analyzer*.

A **logic analyzer** is a small piece of hardware with several digital inputs that can sense whether these inputs are high or low, with extremely high temporal resolution. You can connect these inputs to the **SCL**, **MOSI**, and **MISO** lines of an SPI interface—for example—to see exactly what is happening on these lines. Logic analyzers connect to your PC via USB and work with special software that can decode the raw signal into the bits and bytes that are transmitted across the interface. In the case of SPI, you need to tell the software which input is connected to SCL, to MOSI, and to MISO, respectively. Based on that, the SPI **decoder** will show which data was transmitted.

Figure 7.2 – A generic and inexpensive 8-channel digital signal analyzer

You can obtain signal analyzers for as little as $10, and there is great free **open-source software** (**OSS**) such as the **sigrok** project to use them (`https://sigrok.org/wiki/Main_Page`). *Figure 7.3* shows a screenshot of *PulseView*, the GUI for sigrok, decoding an SPI transaction between an Arduino and the MCP3008 ADC, which we looked at in detail in *Chapter 2, Making Robots Perceive the World with Sensors*:

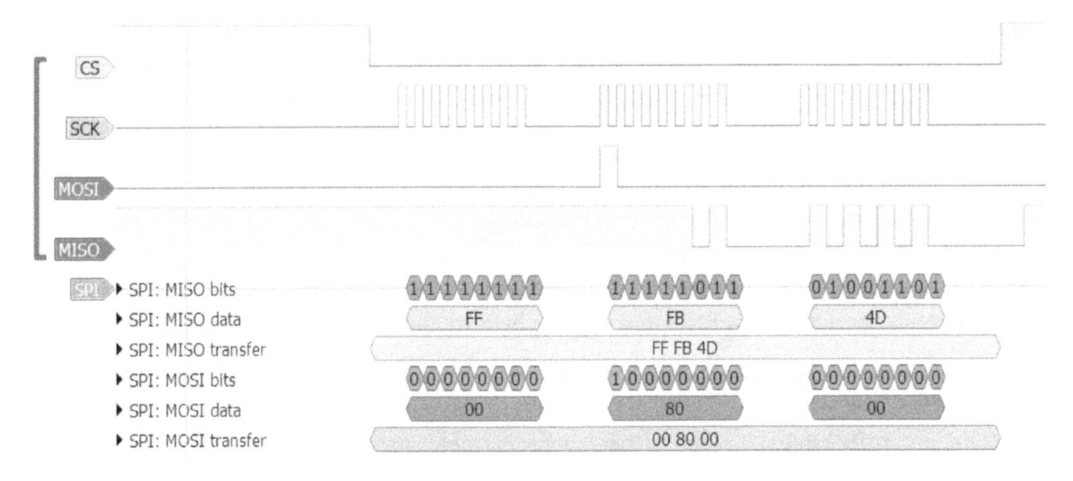


Figure 7.3 – Screenshot of a decoded SPI transaction in PulseView

In this screenshot, you can see that four signals were captured: **Chip Select** (**CS**), SCK, MOSI, and MISO. For each of these signals, you can see their logic levels (high or low) as a continuous trace.

The screenshot also shows the output of the SPI protocol decoder based on the sampled signals. You can see the individual bits and the three corresponding bytes transmitted from the Arduino to the MCP3008 on the MOSI line, and the resulting bits and bytes that are simultaneously transmitted from the MCP3008 back to the Arduino on the MISO line. If you recall the communication protocol of the MCP3008, you can see that the Arduino is requesting the level of the single-ended analog voltage on the MCP3008's channel 0, and the MCP reports back the 10-bit binary value 1101001101, or 845. With a reference voltage of 5V, this corresponds to a measured voltage of 4.13V.

sigrok has built-in protocol decoders for all the communication protocols that we have discussed so far (and many more), making it easy to see exactly which data is being transmitted. This level of insight can be incredibly useful when debugging communications issues, as it allows you to compare what is happening to what the datasheet of the component you are using requires (*Figure 7.4*).

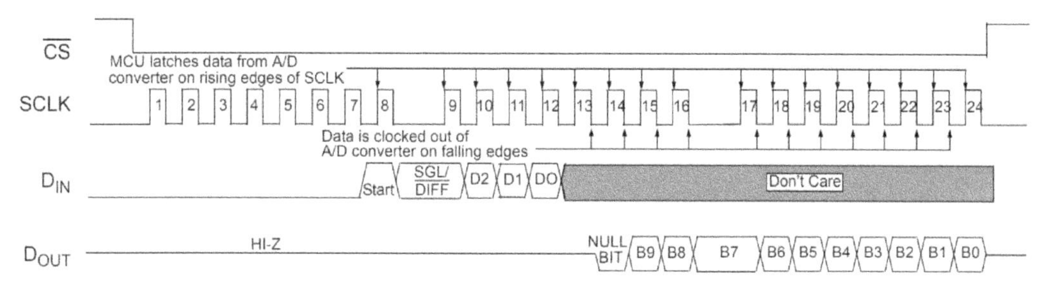


Figure 7.4 – The MCP3008's communication protocol as shown in the datasheet

Compare, for example, the captured communication from *Figure 7.3* with the excerpt from the MCP3008's datasheet in *Figure 7.4*. By comparing the square waveforms of the signal, we can see how the actual transaction matches the description in the datasheet, giving us a lot of confidence that we implemented it correctly.

# Writing testable robot programs

Gaining insight into the inner workings of the program that is running on your Arduino with your own senses is even more difficult than debugging the electronic systems of your robot. That is why it is important to design and write your robot programs with testability in mind from the very beginning.

The most important aspect of testability is modularity. This lets you run a smaller program that only contains some parts of the overall program and test these parts independently. In the previous chapters, we have already learned important techniques for modularity—namely, cooperative multitasking and **object-oriented programming** (**OOP**). If your program consists of multiple cooperative tasks running at the same time, you can simply disable all but one task to only test this one. And if you have a class that can perform a certain core function, such as the `Blinker` class we developed in *Chapter 6, Understanding Object-Oriented Programming and Creating Arduino Libraries*, you only need to test this class once. As soon as you are confident that it works as you expect it to, you can use it as a reliable building block in all your future programs to perform the same function.

It takes some upfront effort and design consideration to write testable, modular programs. For example, if you write a controller to control the position of your robot based on distance sensor input, this controller should only implement the control algorithm. Sensor sampling and motor control should happen in separate functions. This way, the controller can be its own, standalone module that takes in data (that can come from a distance sensor) and outputs a control signal (that can be used to control a motor). In this case, you can use a different signal—for example, from a potentiometer—to generate a test input and observe the output in the Arduino IDE's Serial Plotter for tuning and debugging, even before you have the sensor and motor on hand. Only the next step would be to use the actual sensor data that you have separately tested as input, again while observing the output in the plotter. And finally, you can use the controller to drive your robot's motors, which you will also have already tested separately.

You should aim to make each logical function block of your code—such as a sensor driver, a control algorithm, and a motor driver—its own, self-contained **module** with well-defined inputs and outputs. This will not only make it easier and faster to test but also easier to reuse in a future project. This is especially true if you encapsulate each module as a class and turn it into an Arduino library.

# Using the Arduino IDE for testing and debugging

There are several methods of testing the modules of your robot program. If you write a piece of code that can make an LED blink with variable frequency, you can create the desired frequency with a potentiometer and use an LED to see if the blinker code is reacting correctly to the input signal. However, the Arduino IDE provides us with two additional powerful tools for testing and debugging: the **Serial Monitor** and the **Serial Plotter**. Although we have used them in previous chapters already, let us take a deeper dive into both and learn how we can use them as very effective tools in our development process.

## Using the Serial Monitor as input

The Arduino IDE's Serial Monitor is a character-based interface to send and receive data to and from a connected Arduino via its Serial interface. So far, we have only used it to receive and display data to show the output of the program that runs on your Arduino. But it can also be used to *send data to the Arduino*, which is a very flexible way to input test signals and change parameters while the Arduino is running.

When you open the Serial Monitor, you will notice an empty line on the top. This is where you can enter characters or words that you want to send to your Arduino. They will be sent once you press *Enter*. Next to it, there is a drop-down menu that lets you select a line-ending character that will be automatically added at the very end of your message when it is sent. The **newline** or **end-of-line** (**EOL**) character ('`\n`', ASCII `0x0A`) is the most common line ending, and it is the one that we will be using, too. The other options are no line ending at all, the carriage return character ('`\r`', ASCII `0x0D`), or both newline and carriage return.

Let us write a simple program to demonstrate how to receive and process data coming from the Serial Monitor. This little program is based on the minimal example we wrote for our `Blinker` library, with a few additions:

```
// Include your custom Blinker library.
#include "Blinker.h"

Blinker myBlinker(13, 50);

void setup() {
  myBlinker.begin();
  Serial.begin(115200);
  Serial.setTimeout(0);
}

void loop() {
  myBlinker.blink();
  parse_interval();
}

// Read Serial input and interpret it as blink interval.
void parse_interval() {
  int new_interval = 0;
  if (Serial.available() > 0) {
    // Wait to receive the entire message.
    delay(10);
    if (isDigit(Serial.peek())) {
      new_interval = Serial.parseInt();
      Serial.print("Setting blink interval to ");
      Serial.println(new_interval);
      myBlinker.set_blink_interval(new_interval);
    } else {
      Serial.println("Warning: Invalid data received.");
    }
    while (Serial.available() > 0) {
      // Flush the serial input buffer.
      Serial.read();
```

```
        }
    }
}
```

In this program, we define a new helper function, `parse_interval()`. It first checks if any serial data was received. If there is serial data available, we let it wait for 10 **milliseconds** (**ms**) to make sure that the entire message was received. We then check if the first character of the message is a digit by calling the `isDigit()` function on the returned value of the `Serial.peek()` function. `Serial.peek()` returns the first character in the serial input buffer without deleting it. If it is a digit, we parse the message to an integer using the `Serial.parseInt()` function. The program then confirms the received value back to the Serial Monitor with two calls to `Serial.println()` and sets the blinking interval to the received value.

If the first character is not a digit, this means that the message is invalid. In that case, we simply print a warning message and do not change the blink interval.

Lastly, the `parse_interval` function program discards everything that is left in the input buffer by reading (and thereby deleting) all available bytes. This is at a minimum the newline character and all other characters that might have been trailing the integer value or the invalid first character.

If you load this code onto your Arduino and run it, the LED will initially blink with the 50 ms interval that we specify in the call to the constructor of `myBlinker`. Now, you can open the Serial Monitor on your PC and type a new interval into the input line—for example, the number `500`—and hit *Enter*. You will see a message confirming that your Arduino received the command, and the LED blinking will adjust accordingly, as shown in *Figure 7.5*. If you instead type a number followed by letters, such as `200abc`, the result will be the same. However, if you send a negative number or start the command with a letter—for example, `abc100`—you will see the warning message and the blink interval will not change:

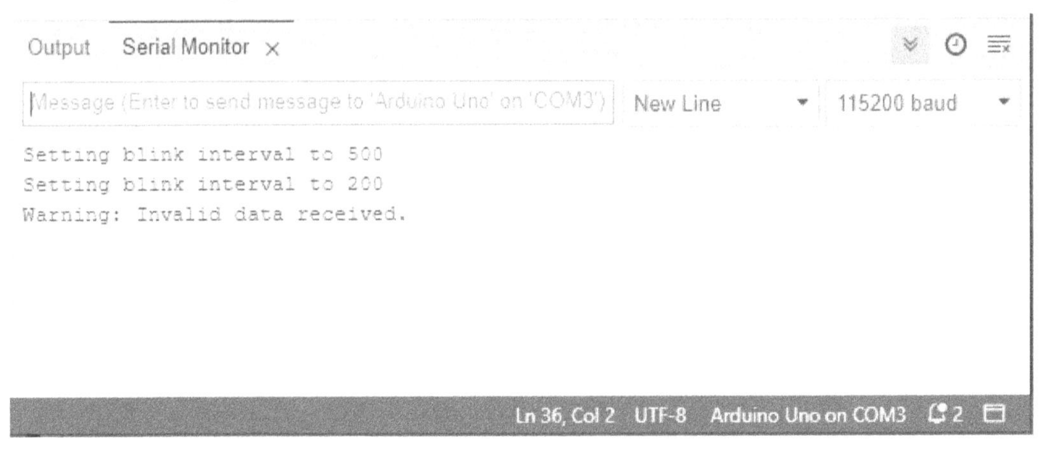


Figure 7.5 – Example output of the Serial Monitor

This code uses the `delay()` function to ensure that the entire message is received before it is processed. Even though it only halts for 10 ms, this makes it a **blocking** function and incompatible with the cooperative multitasking concept. To avoid this problem, we can implement `parse_interval()` as a state machine with the two states of `WAITING_FOR_DATA` and `PROCESSING_DATA`. This is a great exercise to test your understanding of state machines and your ability to translate blocking prototype code into non-blocking code. Before you look at the following code example, take this opportunity to try to implement a non-blocking version of the `parse_interval()` function yourself.

First, we declare and initialize the state and variables of the serial parser state machine:

```
enum {
  WAITING_FOR_DATA,
  PROCESSING_DATA
};
int parser_state = WAITING_FOR_DATA;
long last_rx_data_timestamp = 0;
int rx_wait_time_ms = 10;
```

With these new variables in place, this is what a non-blocking version of `parse_interval()` looks like:

```
void parse_interval() {
  switch (parser_state) {
    case WAITING_FOR_DATA:
      if (Serial.available() > 0) {
        // Data received.
        last_rx_data_timestamp = millis();
        parser_state = PROCESSING_DATA;
      }
      break;
    case PROCESSING_DATA:
      if (millis() - last_rx_data_timestamp >= rx_wait_time_ms)
{
        // Enough time has passed since the first byte was
received.
        if (isDigit(Serial.peek())) {
          int new_interval = Serial.parseInt();
          Serial.print("Setting blink interval to ");
          Serial.println(new_interval);
```

```
          myBlinker.set_blink_interval(new_interval);
        } else {
          Serial.println("Warning: Invalid data received.");
        }
        while (Serial.available() > 0) {
          Serial.read();
        }
        // Reset parser state.
        parser_state = WAITING_FOR_DATA;
      }
    }
  }
```

After having read the previous chapters, all of these changes should look familiar to you. We have split the function into two states. When it is in the `WAITING_FOR_DATA` state and data is received, it saves a timestamp and transitions to the second `PROCESSING_DATA` state. Instead of waiting in a blocking manner, this state checks if sufficient time has passed each time the function is called, and only then processes the available data. As long as only this task runs on your Arduino, the behavior is the same as that of the first, blocking version. However, in a more complex program, this non-blocking version nicely cooperates with other non-blocking tasks, whereas the first version could cause all sorts of difficult-to-debug problems due to its call to `delay()`.

What makes the Serial Monitor such a powerful input tool is that we can use it to input a variety of parameters, not just one. Imagine, for example, we had three LEDs connected to the Arduino, controlled by three instances of the `Blinker` class: `blinker_a`, `blinker_b`, and `blinker_c`. If we wanted to change the interval of all three individually, we could add a key as the first character to our message and use the numerical value according to this key. An implementation of this concept, in particular the `PROCESSING_DATA` state, would look like this:

```
case PROCESSING_DATA:
  if (millis() - last_rx_data_timestamp >= rx_wait_time_ms) {
    // First character is the command key.
    char key = Serial.read();
    if (isDigit(Serial.peek())) {
      // After the key comes the parameter value.
      int value = Serial.parseInt();
      // Decide what to do with the value based on the key.
      switch (key) {
        case 'a':
```

```
            Serial.print("Setting blink interval a to ");
            Serial.println(value);
            blinker_a.set_blink_interval(value);
            break;
          case 'b':
            Serial.print("Setting blink interval b to ");
            Serial.println(value);
            blinker_b.set_blink_interval(value);
            break;
          case 'c':
            Serial.print("Setting blink interval c to ");
            Serial.println(value);
            blinker_c.set_blink_interval(value);
            break;
          default:
            Serial.println("Warning: Invalid key received.");
        }
      } else {
        Serial.println("Warning: Invalid value received.");
      }
      while (Serial.available() > 0) {
        Serial.read();
      }
      parser_state = WAITING_FOR_DATA;
    }
```

Here, we use a `switch` statement to decide which blinker's interval to change, based on the key character at the beginning of the message. If there is no action associated with the key that was received, nothing is changed, and a warning is printed.

This technique can also be extremely useful during controller tuning, as it allows you to change controller parameters on the fly without having to recompile and upload the program for every change. In this case, you might want to transmit floating-point values. You can parse them by using the `Serial.parseFloat()` Arduino function, which otherwise works the same as `Serial.parseInt()`.

## Using the Serial Plotter for live visualization

The Arduino IDE's Serial Plotter is a great tool for visualizing the data produced by the Arduino, be it sensor data, controller output, or the result of intermediate calculations. We have already used the Serial Plotter in previous chapters to display the output of a sensor and to gain insights into the behavior of the position controller that we implemented for our DC servo motor in *Chapter 5, Getting Started with Robot Programming.*

To plot data with the Serial Plotter, we simply use `Serial.println()` to print a value. The plotter will read and plot the serial data one line at a time.

A line can have more than one value. If these values are separated by a comma, a space, or a tab, they are plotted on different curves and in different colors, as you can see in *Figure 7.7*. Using a tab character ('`\t`') as a delimiter is preferable to make the output more human-readable when viewed in the Serial Monitor instead of the Serial Plotter (see *Figure 7.6*). Keep in mind that the newline character (which is automatically added by `Serial.println()`) must only be added at the very end of the line. If a line consists of multiple values and delimiters, all but the last print must be done with `Serial.print()` and not `Serial.println()`.

The Serial Plotter supports a color legend that you can add to your plot. To enable this feature, each value must be preceded by its label and a colon.

By default, the Serial Plotter constantly adjusts the scaling of the vertical axis to fit the data that is currently visible in the plot window. You can defeat this feature by printing two separate, horizontal data lines containing the constant minimum and the maximum values that your data will contain. That way, the plot will always be scaled around these two values.

The following code example uses all these features of the Serial Plotter. At every iteration of the `loop()` function, this program computes the value of time-based sine and cosine waves with fixed frequencies and amplitudes. It then uses seven `print` statements, the last of which is `Serial.println()`, to print a line with four data points: the value of the sine function, the product of the values of sine and cosine, and the minimum and the maximum of this function to prevent auto-scaling. The labels for these four data lines are written directly preceding the values. Every iteration of `loop()` ends with a call to the `delay()` function. The longer this wait time, the slower the plot will progress in the Serial Plotter:

```
const float sin_freq = 2.0;
const float cos_freq = 4.0;
const float sin_amp = 10;
const float cos_amp = 2;

void setup() {
  Serial.begin(115200);
}

void loop() {
  // Argument for 1Hz sin and cos waves.
```

```
  float arg = 2 * PI * millis() / 1000.0;
  // Sine and cosine waves.
  float sinwave = sin_amp * sin(sin_freq * arg);
  float coswave = cos_amp * cos(cos_freq * arg);
  // Print four columns of data.
  Serial.print("sin:");
  Serial.print(sinwave);
  Serial.print('\t');
  Serial.print("sin*cos:");
  Serial.print(sinwave * coswave);
  Serial.print('\t');
  Serial.print("min:");
  Serial.print(-sin_amp * cos_amp);
  Serial.print('\t');
  Serial.print("max:");
  Serial.println(sin_amp * cos_amp);
  delay(10);
}
```

In the following two screenshots (*Figure 7.6* and *Figure 7.7*), you can see snapshots of the output of this program both in the Serial Monitor and in the Serial Plotter:

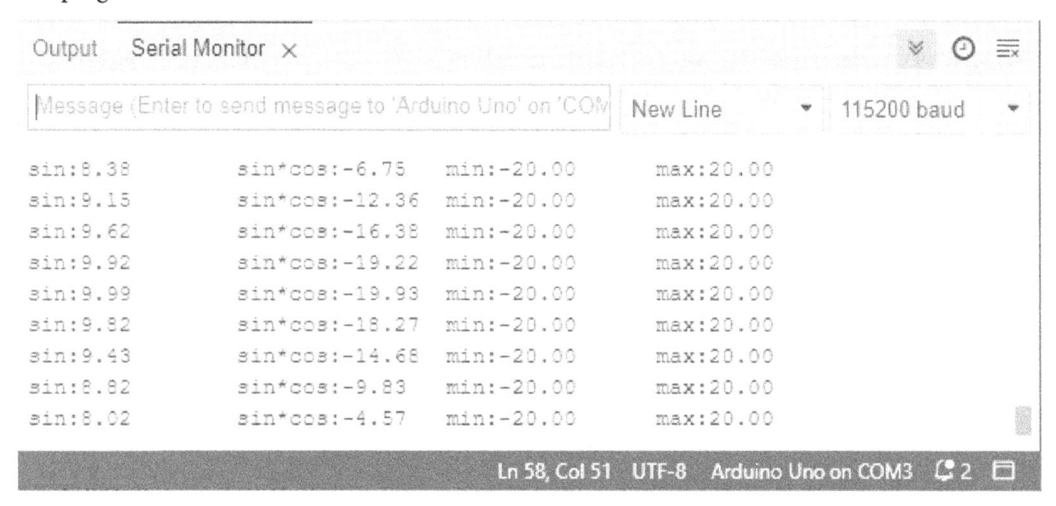


Figure 7.6 – Output of the plotter demonstration code in the Serial Monitor

As you can see, the values are well aligned, and the columns are easy to read thanks to the use of the tab character as a delimiter between the individual data points:

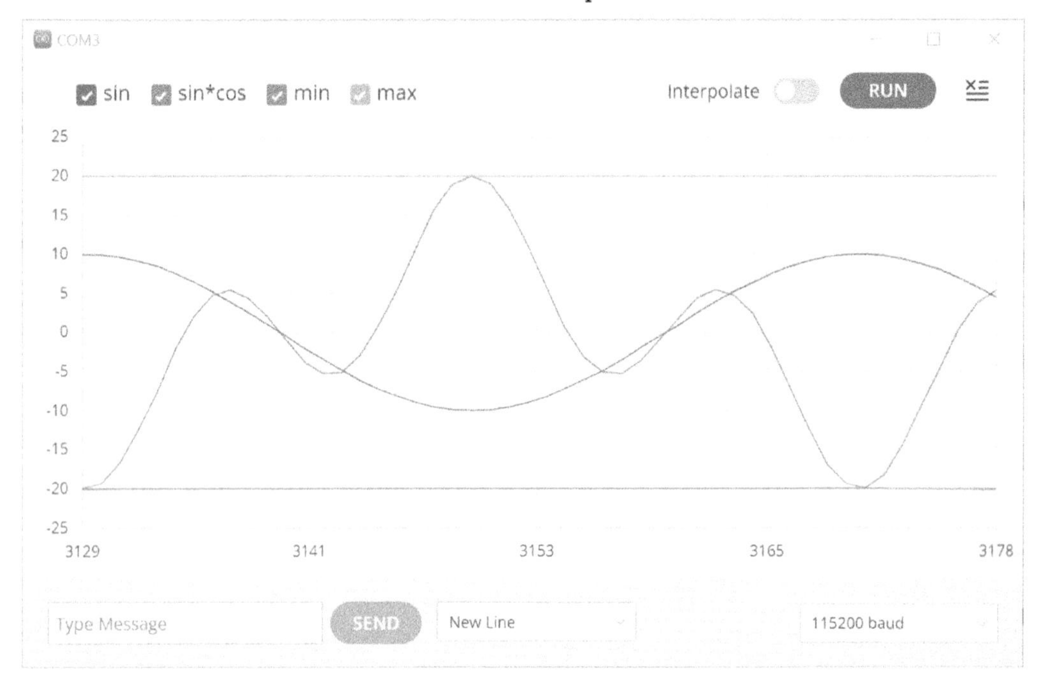


Figure 7.7 – Output visualized in the Serial Plotter

In this screenshot, you can see the two data curves in red and blue, with the corresponding entries in the legend. You can also see the minimum line in green and the maximum line in yellow that prevent the plot from auto-scaling. You can disable each of these curves by unchecking their checkmarks in the legend at the top left of the window.

The plotter window further has three control elements in the top-right corner:

- The **Interpolate** switch enables spline interpolation to make the curves look smoother. Be careful when using this feature, as it can produce misleading plots.
- The **STOP/RUN** button is a very useful feature to take snapshots of a plot, as for this book. Pressing **STOP** instantly freezes the plot, and pressing **RUN** lets it resume.
- The button that looks like several lines and the letter **x** clears the data that is currently being displayed.

You can familiarize yourself more with the Serial Plotter by playing with the parameters at the beginning of this code. You can examine what effect enabling auto-scaling has (by omitting the **min** and **max** data lines) and what effects changing the print period has. You will find that too short of a print period can lead to unexpected, undesirable effects.

**Explanation**

`Serial.print()` writes the data to a transmit buffer and returns immediately. Behind the scenes, an interrupt-driven mechanism feeds the data from this buffer to the UART peripheral, byte by byte, without using the CPU. However, if you call `Serial.print()` too frequently or print too much at once, you may fill the transmit buffer faster than the UART can send the data. In this case, `Serial.print()` needs to wait for the UART to transmit the data and free up space in the buffer. `Serial.print()` then becomes a blocking call that can alter the behavior of your program. For this reason, adding printouts for debugging can sometimes make the bug disappear by unintentionally changing the timing of your program. This problematic phenomenon is jokingly called **load-bearing printf**. In practice, you want a high baud rate and leave a few tens of milliseconds between calls to `Serial.print()` to avoid this problem. You can also programmatically check the available space in the transmit buffer with the `Serial.availableForWrite()` function before a `print` command to be sure that you are not clogging up the buffer.

The Serial Plotter also has an input line, just like the Serial Monitor. With some modifications based on what we learned in the previous section, you can write a program that allows you to change the parameters of the sine and cosine waves on the fly right from the Serial Plotter. This is a great exercise to test and further your understanding of this chapter.

# Summary

In this chapter, we discussed the importance of component testing and how to design the hardware and software components of your robots with testability in mind. It is crucial to split your systems into modules that can be tested individually before you combine them into a more complex system that can be difficult and frustrating to debug. We further learned how to use the Arduino IDE's built-in Serial Monitor as a flexible tool during the testing phase of your project, allowing you to input arbitrary values and change any parameter of your program while it is running. Finally, we took a closer look at the Arduino IDE's immensely useful Serial Plotter and discussed its features. This chapter concludes *Part 2, Writing Effective and Reliable Robot Programs for Arduino,* of this book, in which we learned a lot about writing effective, high-quality code for Arduino robots. Now that we know the components we can choose for our robot and how to write the code to use them, the next part of this book will address many practical matters that come up when building a real robot. The next chapter will explore important aspects of the mechanical design of your robot, including material selection, CAD, and 3D printing.

## Further reading

Here are a few additional resources that can help you learn even more about debugging your robot:

- An in-depth guide on using digital multimeters for debugging electronics: `https://learn.sparkfun.com/tutorials/how-to-use-a-multimeter/all`
- A great tutorial on getting started with logic analyzers and sigrok PulseView: `https://learn.sparkfun.com/tutorials/using-the-usb-logic-analyzer-with-sigrok-pulseview`
- The official how-to guide for the Arduino Serial Monitor: `https://docs.arduino.cc/software/ide-v2/tutorials/ide-v2-serial-monitor`
- The official how-to guide for the Arduino Serial Plotter: `https://docs.arduino.cc/software/ide-v2/tutorials/ide-v2-serial-plotter`

# Part 3: Building the Hardware, Electronics, and UI of Your Robot

This part focuses on showing you how to build mobile robots completely from scratch using basic materials and tools, modern CAD, and 3D printing. It also covers the fundamentals of laying out the electrical power system of your robots, which distribute the power from the batteries to all their components safely. In this part, we also cover UI elements such as light, sound, and displays to make your robots more interactive. And finally, you will learn how to add wireless interfaces to your robot, for example, to control it from your phone using Bluetooth.

This part has the following chapters:

# 8

# Exploring Mechanical Design and the 3D Printing Toolchain

This chapter is a structured overview of mechanical design principles and important considerations when designing an Arduino robot. It covers design choices, material choices, and the trade-offs between using off-the-shelf components versus robot design from scratch with **computer-aided design** (**CAD**) and 3D printing.

The chapter is structured as follows:

- Principles of mechanical design for DIY robots
- Exploring materials and manufacturing techniques
- CAD
- 3D printing

## Technical requirements

This chapter is purely informational; there is nothing that you need to follow along. We will cover principles, materials, methods, and 3D printing. By the end of the chapter, you will hopefully want to start learning CAD, dive into the world of 3D printing, or head to the hardware store, pick up some materials, and start building robots.

## Principles of mechanical design for DIY robots

In previous chapters, we have taken in-depth looks at the core components of any Arduino robot: the sensors, the actuators, and the Arduino board itself. Once you have selected all these components and have written and tested the Arduino program to use them on a simple benchtop setup, it is time to build your robot! The mechanical components of your robot are what hold all the other parts together, and they need to be sufficiently stiff, durable, and reliable to make your robot project a success. Depending on the kind of robot you are building, the requirements for the mechanical components can vary quite a bit. For a combat robot, your material of choice would be metal to make it as durable as possible, even though metal is expensive and difficult to work with. For a flying robot, your focus would be on a lightweight design with low-density materials to minimize power consumption and maximize flight time.

The design considerations we discuss in this chapter are inspired by a low-cost, wheeled mobile robot that is well suited for many kinds of experiments. The requirements for the mechanical design for this use case are as follows:

- Low cost
- Easy to build
- Low part count
- Easy to modify
- Durable

Let us dive in and learn how to build robots that meet these requirements.

## Keep it simple

The most important guideline to building your own robots is to keep it as simple as possible. The simplest design that works for your robot is usually the best. If your robot is going to drive over flat ground, a three-wheeled design with two driven wheels and a passive caster wheel is going to be the best choice. This type of drive system has the smallest part count and the highest efficiency. Using four or more wheels, or tracks, would make the design unnecessarily complicated. If your robot needs to pick up little boxes, a simple jaw gripper with a single joint is a better choice than a humanoid hand with many motors and joints.

It is quite common to see inexperienced robot builders choosing design ideas that are overly complicated. Consequently, the entire robot project can suffer or even fail due to mechanical issues with a component that would not have been necessary for the core function of the robot in the first place. It can be tempting to start building your initial ideas right away, but it is important to try to simplify the design as much as possible to give yourself the best chances of success. Time at the drawing board trying to simplify the robot design is very well spent—much better than frustrating hours of fixing an overly complicated design that does not work reliably.

The easiest way to keep your robot design simple is to use an off-the-shelf robot kit that already comes with all the hardware you need, such as motors and wheels, a battery tray, attachment points for sensors, and maybe even a gripper. There is a very large variety of kits such as these available, and they can be a great way to get started. You might not find the exact right kit for your project, but adding additional functionality such as sensors, arms, or grippers to a sturdy and proven platform can still be the easiest, quickest, and most cost-effective route to bring your project idea to life.

If you do not have access to a kit that fits the needs of your robot projects, aim to build your robot from as few components as possible. The best way to achieve this is to choose components that encapsulate as much functionality as possible. Let us look at the challenge of driving one of the wheels of your robot, for example. To drive the wheel, you will need an electric motor, a reduction gearbox, rotary bearings to carry the weight of the robot, a mechanical structure to hold everything together, and—ideally—an encoder that lets your Arduino keep track of the wheel position. One option is to obtain all these components separately, design and build the mechanical structure, and assemble everything yourself. This process can be fun and educational but can also cost a lot of time and money, and it can easily lead to frustration. The better option in the spirit of keeping things simple is to choose a ready-to-use geared DC motor with an integrated encoder, such as the one that we used to implement our DC servo motor example. Unless you have highly specific requirements, a geared DC motor that fits your needs is probably readily available—for example, from *Pololu* (`https://www.pololu.com/`). *Figure 8.1* shows a geared DC motor and an RC servo motor, components that are often sensible choices for powering DIY Arduino robots:

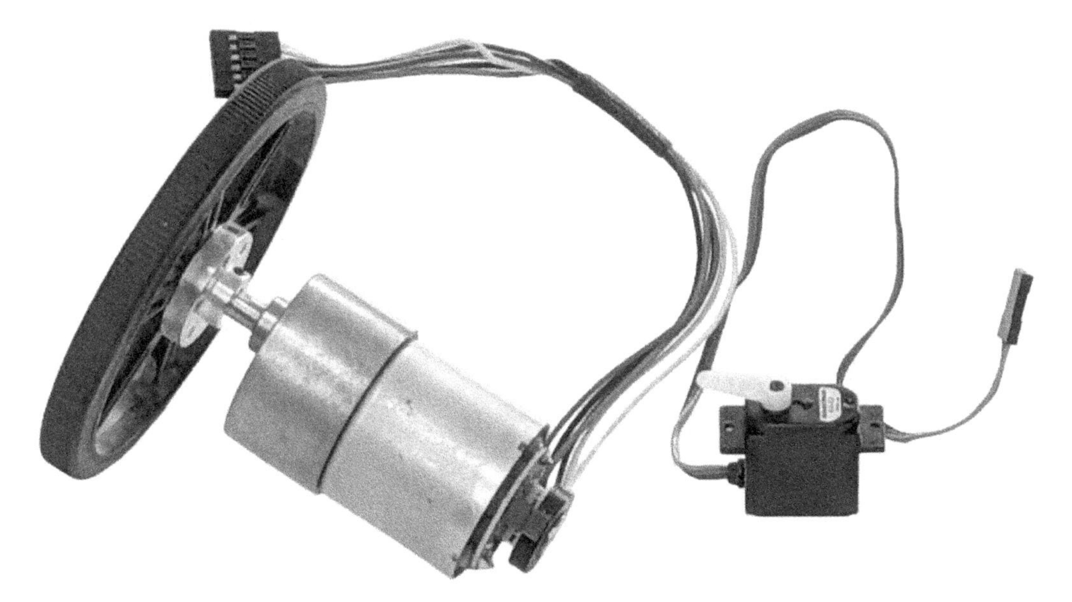

Figure 8.1 – Integrated components such as geared motors with encoders (left) or RC servos (right) are an excellent way to keep your design simple and modular

This approach works well for all aspects of your robot, including mechanisms, sensors, and actuators. Before you spend time building a subsystem such as a motorized camera gimbal or a distance sensor from several discrete components, try to find out if something like it is already available as an integrated component. This can dramatically speed up the development of your robots and increase the chances that your project will become a success. It can potentially reduce the cost of your robot, too.

### Make it modular

Just as we like to write Arduino software in a modular way that allows us to test subcomponents independently, it is also a great idea to make the mechanical design as modular as possible. If your robot contains several mechanisms, try to keep them as separate as possible and develop and test each one individually. For example, if you are building a two-wheeled, self-balancing robot with two arms, grippers, and a moving head, it is a smart approach to work on these subcomponents individually first before integrating them all into one robot. This way, you can test the arms on the benchtop, then the grippers, and then the neck mechanism without having to fear damage to them while you develop and test the two-wheeled base of your robot. Integrating mechanical components that are tested and work individually is much easier than integrating first, and then testing everything all at once.

### Make a plan

It is important to make as detailed a plan as possible before you start building your robot. Try to think of all the components that need to go into your robot, how much space they take up, how much they weigh, and if they need cooling or have other specific requirements. And if possible, leave some extra room for wiring and other additions that you might not think about at the time of designing. Otherwise, you might have to make design modifications during your build or after the first assembly test, which costs time and is likely to make the final design of your robot more complicated. A good plan will also allow you to order all the parts you need at once, saving time and money on shipping if you buy them online.

It often takes quite a bit of experience to make a good plan for your robot design ahead of time. After you have built a few robots, you will find it easier to think of all the details, and you will know many techniques that help you to make do with what you have on hand, even if you did not order all the parts and materials you need. The more robots you build, the more tools and leftover parts you will have, making it easier to complete a project even if you decide to deviate from your initial plan a little bit.

## Exploring materials and manufacturing techniques

For a successful start in the world of DIY robotics, it is a good idea to base your first own Arduino robot on a robot kit that comes with most or even all mechanical components you need, such as motors, wheels, and a chassis. But soon, you will want to add more components to this first robot until eventually, you want to build your own robots completely from scratch. With a small, wheeled mobile robot in mind, let us look at the materials and techniques that are useful to know when progressing from ready-to-use kits to building robots from scratch.

As soon as you start adding components to your robot, you need to securely connect them to the existing chassis. Mechanical connections can either be permanent or temporary. A permanent connection is one that is not meant to come apart anytime soon—for example, when you connect a robotic arm to your mobile rover. A temporary connection is meant to come apart easily—for example, to attach a removable battery or a cover that needs to be removed for accessing the battery.

## Permanent connections

The most common permanent connections between parts are formed by screws and bolts. Screws have a tapered, pointy end and a coarse thread. They can cut a mating thread into the material they are screwed into, making them extremely easy to install and very versatile. Screws are a fantastic way to attach things to soft materials, such as wood and many plastics. In contrast, they do not work well for hard materials such as aluminum and other metals and do not provide much strength in very soft materials such as foam board. You might be able to drive a screw into the material without drilling a pilot hole first, but unless you are working with extremely small screws or soft material, it is good practice to drill a pilot hole to avoid cracking or material deformation. To attach a part to a structure with a screw, the part needs a hole that is large enough for the screw to loosely fit through. Mark the position of the hole on the supporting structure and drill a pilot hole that is not bigger than the core diameter of the screw. Then, align the part and the structure again, insert the screw, and carefully tighten it with a screwdriver. If you tighten the screw too much, you run the risk of stripping the thread in the support, so be careful not to overtighten it.

Bolts are similar to screws, but they do not have a tapered end and have a much finer thread. They are not meant to cut a mating thread into the material themselves but to be screwed into a nut or a pre-cut thread in a hard material such as aluminum or steel. To connect to parts with a bolt and nut, both parts should need aligning holes with the same diameter. Then, you can insert a matching bolt through the holes and clamp the two parts together by putting on and tightening a nut from the other side. To avoid the nut loosening due to vibration, it is best to use locking nuts wherever possible. These have a little plastic ring insert that grips the bolt and prevents the nut from coming loose. Bolts and nuts can give much stronger connections than screws and are the method of choice to connect hard materials, such as two aluminum parts. Both types of connections are illustrated in *Figure 8.2*:

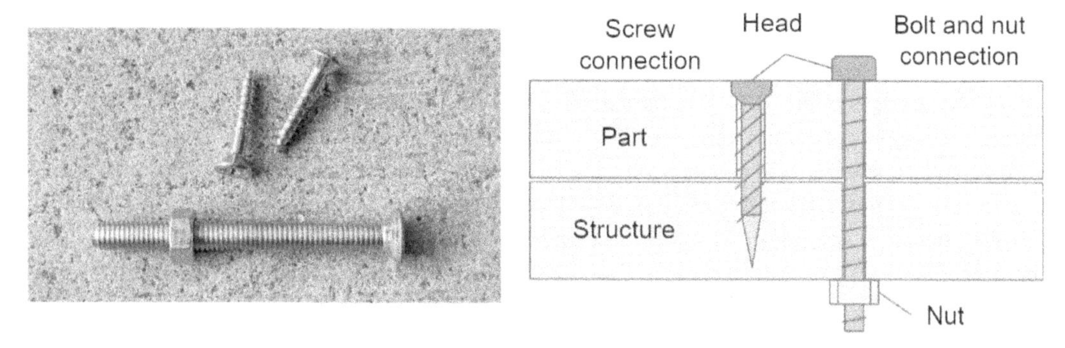


Figure 8.2 – Two screws and a bolt with a nut on the left/depiction of the different connection types on the right; notice the coarse threads and the tapered ends of the screw

Threaded rods are essentially very long bolts without a head. You can cut them to length with a saw and, with a nut on either end, use them to clamp together a thick stack of material. You can also use threaded rods and several nuts to build a stack of multiple levels of sheet material with space in between, each of which can hold a variety of components such as batteries, motor drivers, or sensors. This is a common technique to build very compact robots.

Another common way to permanently connect parts together is by gluing them. Even though gluing might seem like a simple option, it can often be a source of trouble for the mechanical integrity of a robot. That is because there are a lot of materials and glues, environmental conditions, processes, contaminations, and use cases that all affect the choice of glue and the quality of the glue joint. One of the most useful classes of glues is *cyanoacrylate-based superglue*, commonly referred to as CA. These glues quickly and permanently bond together a broad range of materials. CA glue joints tend to become brittle over time, so while they are great for prototyping and quick repairs, they are not the perfect choice for building your robot. They also produce irritating fumes and permanently bond skin to skin, so be sure to only use them in well-ventilated spaces and handle them with great care. Another useful class of glue is *two-component epoxy*. The two components are a synthetic resin and a hardener agent. After mixing the two, the resin polymerizes and turns into a durable plastic, adhering very strongly to most materials. These glues are commonly sold in a double syringe, making it easy to mix them in the right ratio. They are not as easy to use as CA glues, but they can fill gaps and work great on foam or porous materials where CA glues typically fail. There are many other types of glue for all kinds of material pairings, and it is certainly worth experimenting with different glues to find the best for your application. The quality and strength of a glue joint depend a lot on the process, so it is always a good idea to read the instructions for your specific glue. They might even vary depending on the material that you are bonding together, even if you are using the same glue. Most glue joints require some amount of surface preparation. By far the most crucial step of preparation is to clean the surfaces before gluing them together. Residual dust, oil, or grease can dramatically degrade the performance of a glue joint.

Yet another way to permanently join two parts together is with double-sided tape. The most useful type of tape for DIY robotics has a strong adhesive on both sides and a thin layer of foam substrate in between. The foam allows it to conform to surfaces even if they are not perfectly flat, which leads to very good bonds even in challenging situations. The foam also provides some amount of vibration damping and shock insolation. Double-sided tape can therefore be a good way to permanently mount electronic components such as a battery box, a sensor PCB, or even a small RC servo motor to your robot. In contrast to screws and bolts, double-sided tape can be difficult to remove and can leave a sticky residue, so be sure to only use this type of tape when you are sure that the connection will be permanent.

Cable ties, also called zip ties, are common, inexpensive, and incredibly useful devices to connect not only *cables* to a structure, but also two parts to one another, or a part to a structure. Cable ties are ideal to organize the cabling inside your robot by bundling the cables and attaching them to the structure of your robot. You can also use them as straps to strap a motor to a motor mount or a battery to the robot chassis, for example. Even though they are permanent, they can be easily removed by cutting them. *Figure 8.3* shows various glues, tape and cable ties that you can use to connect components of your robot

Figure 8.3 – A variety of means to form permanent connections between parts: epoxy glue in a syringe, hot glue gun, double-sided tape, and cable ties

Combining double-sided tape with zip ties can make for durable and versatile connections that give you the benefits of both. The tape securely holds the part in place and provides a strong bond even if the surfaces are not perfectly flat or parallel. By reinforcing this connection by additionally strapping the part down with one or more cable ties, it can become strong and stiff enough to even attach motors and gearboxes.

## Temporary connections

Temporary connections typically do not need to be very strong; they just need to hold something in place for a while. Let us look at common ways to achieve this.

Hook-and-loop tape is a combination of two types of self-adhesive tape: hook tape and loop tape. If you want to use this type of tape to create a mechanical connection between two parts that can be easily taken apart anytime, all you need to do is stick the hook tape to one part and the matching loop tape to the other. If you bring them into contact, the hooks will latch onto the loops and form a connection that is strong enough to hold a PCB or even a battery in place. This type of connection tends to be good at isolating vibrations and mechanical shock, which can make it useful even if you do not need to take it apart frequently:

Figure 8.4 – The two sides of self-adhesive hook-and-loop tape (loops on the left, hooks on the right)

It is a good idea to decide on which type of tape you stick to the permeant and which to the removable part of the connection, and consequently follow this convention in all your projects. For example, if you use hook-and-loop tape to attach batteries to your robots, you want all your batteries to have the loop tape and all your robots to have hook tape. This makes it easy to use any of your batteries on any of your robots.

Magnets are useful to form another class of temporary connections. To attach a removable cover, for example, you can use magnets in the cover and either magnetic metal (such as steel screws) or magnets in the base of the robot. If you bring them into close proximity, they attract each other strongly enough to form a mechanical connection. A benefit of magnetic connections is their ability to self-align. The magnets will automatically snap into the position that brings them the closest together, without requiring any other alignment features. You can buy strong and inexpensive neodymium magnets in many shapes and sizes online, and they can be glued into place with CA or epoxy glues. These magnets are made of a brittle material that quickly oxidizes when it is exposed to the moisture of ambient air, so they are typically coated in a silver metallic coating. If this coating is damaged and the core material is exposed, it is best to replace the magnet, as it will quickly deteriorate. If you use two magnets for a connection, one on either side, it is crucial that they face each other with opposite poles to attract one another. If they face each other with the same pole, they will instead repel each other. So, always confirm the correct polarity before gluing magnets into place. One possible side effect of magnets in your robot is interference with magnetometer sensors that are commonly part of integrated orientation sensors. These sensors use the magnetometer as a compass to determine their heading, and they can be fooled by nearby magnets. If your robot relies on a magnetometer and you want to use magnets in your design, you need to space them apart as far as possible from one another and ideally conduct some experiments to determine if there is interference.

At some point, you will want to build your own mechanical components to expand and modify your robots. Out of the wide range of materials to choose from, let us take a look at a small selection that is very useful, cost-effective, and easy to work with.

## Wood

Wood, particularly plywood sheets, can be a great material for building small robots. Plywood is inexpensive and very easy to cut into shape using hand tools such as a fret saw. You can also cut plywood using a laser cutter if you have access to one. You can attach the parts of your robot to a plywood structure with screws, and due to the softness of the plywood, you can usually skip drilling the pilot hole. You can also attach parts with a bolt and nut connection, but you will need washers under the bolt head and under the nut to distribute the pressure and avoid damaging the soft wood. You can also glue wood parts together using specialized wood glue, CA, epoxy, or even simple hot glue from a glue gun.

## Acrylic

Acrylic sheet material is another excellent material to build the chassis or other parts of your robot from. Acrylic is a transparent plastic that is significantly harder and much more brittle than plywood, which means you cannot put a screw into it without drilling a pilot hole. Instead, you can cut threads into acrylic as you would into metal so that you can thread bolts right into the acrylic piece. Because it is harder and more durable, the acrylic sheet material is harder to work with than plywood but still much easier to cut and drill than metal. Acrylic can locally melt and stick to the saw if you use an electric jigsaw, so you need to make sure to use your saw at a low-speed setting. It can also be easily cut with most laser cutters if you have access to one. Acrylic is a thermoplastic, which means that you can form it into shape while it is hot, and it retains this shape when it cools down. A heat gun is all you need to sufficiently heat a sheet of acrylic. This is an extremely useful technique to make curved parts for your robot without having to machine or 3D print them. A big advantage of acrylic over plywood is its durability and resilience to moisture and water. Especially if you build robots that are supposed to operate outdoors, acrylic is the better choice. You can also use acrylic sheets or tubes and glue to build watertight and even airtight containers—for example, to build a robotic submarine.

## Aluminum

Of all the metals you can use to build DIY robots, aluminum is the easiest to work with. It is commonly sold online or in hardware stores as sheet material or extruded profiles, such as tubes, I-beams, or corner brackets. Extruded aluminum profiles can be a fantastic way to add a lot of structural stiffness to your robot by attaching them to parts made from wood, acrylic, or 3D-printed plastic. Aluminum is soft enough that you can still process it with common hand tools and a handheld power tool, and you can cut threads into it using a tap to insert bolts. Besides its high strength and stiffness compared to wood and acrylic, aluminum is also an electrically conductive material and an excellent heat conductor. When using aluminum parts in your robot, make sure that electrical contacts such as the underside of your Arduino board do not touch the bare aluminum and cause potentially damaging short circuits. You can use structural aluminum parts as heat sinks—for example, for the motor controllers of your robot. Some motor controller ICs are meant to be bolted to a heat sink, and a large aluminum part can be excellent for this purpose.

### Other materials

If even aluminum is not strong enough for your design, you can consider using steel parts. However, steel is significantly more difficult to machine, and you might need to have access to a professional machine shop to successfully work with steel. You can also consider carbon fiber-reinforced material, which can be as strong as steel but is much lighter and easier to work with. However, you need to ensure proper ventilation and wear protective equipment when cutting or drilling carbon fiber to protect yourself and others from the microscopic fibers that are released and that can be damaging to your health. Carbon fiber is also an exceptionally expensive choice of material. On the lighter and less strong side is foam board, often sold as insulation material at hardware stores. If your design does require a lot of material strength and compactness, foam board can be a good choice. It is extremely easy to cut and drill into, bonds well with hot glue, epoxy, double-sided tape, or specialized foam glue, and is very cost-effective. It is often the material of choice for flying robots due to its low weight. Keep in mind that foam board is a great thermal insulator, so it might prevent the power components of your robot from receiving adequate cooling and can make them overheat faster. Many materials—including foam board, plywood, and 3D printed parts—can be reinforced by laminating fiberglass sheets on top of them, using epoxy resin. This technique is commonly used for building model aircraft and model rockets. Once you master it, you can build incredibly strong yet still lightweight parts in almost any shape.

## CAD

As we discussed earlier, planning your robot's mechanical structure is an essential part of a successful robot project. Your plan can either consist of a few sketches on paper or a digital 3D design. A digital design, often called CAD, has many advantages. It allows you to iteratively change it, inspect every aspect of it to spot interferences and other problems early on, animate and test mechanisms before you have even built them, and easily share and collaborate on your design with others. It also allows you to export your design as files that are needed for manufacturing, such as CNC machining or—most importantly for DIY robotics—3D printing.

There are several user-friendly CAD software packages available, some of which have free versions for the hobby and maker communities. If you do not have CAD software yet that you are familiar with, Onshape (`https://www.onshape.com/en/products/free`) is an excellent choice for DIY roboticists. It is a state-of-the-art parametric CAD tool that offers a free version for hobbyists and makers. It runs in the cloud, such that all you need is a web browser to use it. It makes collaboration with others and version control of your design extremely easy. Onshape covers all the needs for designing your own Arduino robots and has comprehensive online documentation. Other popular choices include the free, browser-based Tinkercad (`https://www.tinkercad.com/`) and the widely used SolidWorks (`https://www.solidworks.com/`). *Figure 8.5* shows a screenshot of a 3D part modeled in Onshape:

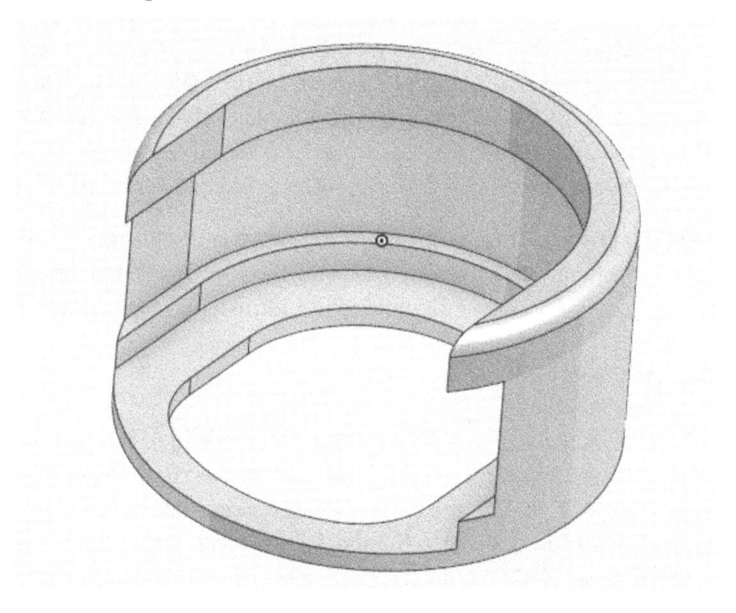

Figure 8.5 – A custom cover part for an optical sensor, 3D modeled in Onshape

An introduction to 3D modeling with CAD software is beyond the scope of this chapter, but there are many great resources available on the internet to get you started. At a high level, CAD allows you to create 3D models of individual parts and then combine these parts into assemblies. Assemblies can have moving joints, so you can see your mechanisms in action and verify that they work as expected. Many vendors of hardware components provide digital 3D models of their products that you can import into your CAD assembly to make sure everything fits, even before you build your robot.

You can export the parts that you designed—for example, as STL files for 3D printing.

# 3D printing

3D printing is a popular method from the wider field of **additive manufacturing** to make a very wide variety of mechanical parts cheaply at home. Mastering the 3D printing toolchain, from the design idea in your head to a physical part in your hands, can lift the mechanical side of all your robot projects to the next level.

## What is a 3D printer?

A 3D printer is a machine that can create arbitrarily shaped 3D parts from an input material. There are many types of 3D printers. We will focus on the **Fused Filament Fabrication** type (or **FFF** for short) since these are by far the most widely used machines. An FFF 3D printer takes in material in the form of a thermoplastic filament from a spool, extrudes it through a hot nozzle where it melts, and deposits the molten material, layer for layer, onto a build plate. The nozzle can move freely in the horizontal *x* and *y* directions, allowing for extremely precise material deposition within a horizontal layer. With each layer, the nozzle moves up a little so that each layer adds to the height of the part.

Even in the class of FFF 3D printers, there are many subtypes that can print varied materials, support structures, and reinforcement materials. They vary in print speed and build volume, and range from finicky but cheap hobby-grade printers to industrial-grade, ultra-reliable machines.

## 3D printers for DIY robotics

For DIY Arduino robots, let us focus on one very specific category of printers: hobby-grade, single-extruder FFF printers. These are by far the most common printers, and you can purchase a very decent one starting at around $250. The most common filament material used by hobbyists is a thermoplastic called **polylactic acid**, or **PLA** for short. PLA has many favorable properties that make it so popular. It melts at relatively low temperatures compared to other materials such as **acrylonitrile butadiene styrene** (**ABS**) or nylon. In contrast to other materials, it does not require a heated print bed or a heated print chamber, all of which allow for simpler and more energy-efficient printer designs. PLA does not give off dangerous fumes during printing, which makes PLA printers perfectly safe to have inside your home, even without an enclosure or fume hood. PLA also comes in a wide variety of colors, is non-toxic, and is cheap compared to other materials. PLA is even biodegradable under industrial composting conditions. While PLA lacks the incredible strength of fiber-reinforced filaments or the high-temperature tolerance of Nylon materials, it easily meets the needs of most DIY robotic projects.

A typical printer of this kind, such as the one shown in *Figure 8.6*, has a build plate for the part to sit on and a printhead that feeds the PLA filament through a heated nozzle to deposit it as the next layer of the print.

Figure 8.6 – A typical hobby-grade FFF printer for PLA material

The printer has at least four stepper motors: two to move the printhead in each horizontal direction ($x$ and $y$) within a layer, one to move the printhead up and down between layers, and one to drive the extruder that pushes the filament through the hot nozzle. Most printers also have a display and some control elements, a USB interface, an SD card slot, and a Wi-Fi interface. To use the printer, you need to transfer a file that contains machine-readable G-code to it via one of its interfaces and select that file for printing.

Popular makes of hobby-grade PLA printers include Prusa, Ender, Ultimaker, and Monoprice.

## Slicing

Slicing is the part of the 3D printing toolchain that converts a 3D model into the **G-code** file needed to instruct the 3D printer how to print the part. The process is called slicing because of the way an FFF printer builds the part layer by layer or slice by slice. A slicer is a piece of computer software with a graphical user interface that lets you control many parameters of the process. An example of widely used and freely available slicing software is Cura by Ultimaker (`https://ultimaker.com/software/ultimaker-cura`). A slicer takes in an STL file, which is a standardized representation of the surface of an object represented by many small triangles. All major CAD programs can export 3D parts as STL models, so you can use any slicer with a CAD program simply by exporting an STL file from your CAD software and importing it into your slicer. There are even third-party services that allow you to upload your STL file and that will print it for you from a variety of materials. These can be great if you do not have your own 3D printer, or if you need a material or part quality that is outside the capabilities of your printer. Examples of these services include Shapeways (`https://www.shapeways.com/`), Protolabs (`https://www.protolabs.com/`), and Xometry (`https://www.xometry.com/`). In your slicer, you will need to specify the parameters of your printer in the slicer, or you can use a predefined parameter set that matches the printer you want to use. You can then set the parameters of the print, such as layer thickness, print speed, nozzle temperature, build bed temperature, infill density, support type, and many more. There are a lot of great resources available on the internet to learn more about 3D printing, but if you have a printer, it can also be a fun way to learn by trial and error. Luckily, PLA is inexpensive, and it can be fascinating to watch your 3D printer build little test parts and study the effects of different settings. Once the STL file is imported and all the settings for your printer and the specific print are dialed in, the slicer will generate a G-code file that you can transfer to the printer. *Figure 8.5* shows a visualization of the slicer output. You can see the paths that the printer nozzle will take to create this part.

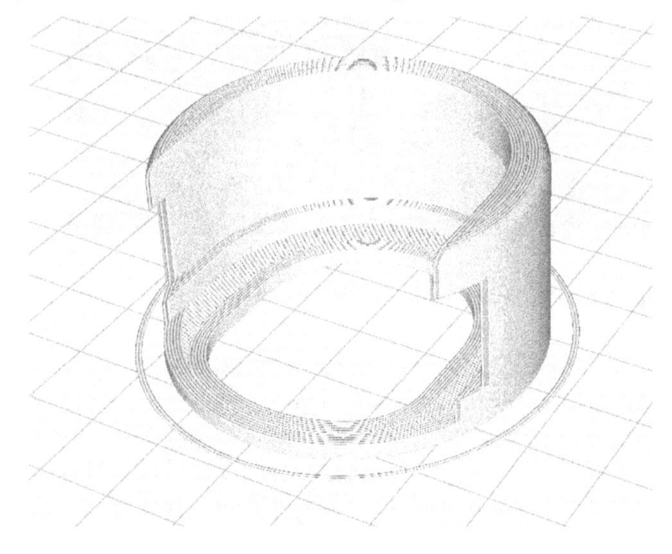

Figure 8.7 – The same cover part as in Figure 8.5, sliced and previewed in Ultimaker's Cura software; notice the purge line around the part

For hobby-grade printers or printers that you build yourself, finding the set of parameters that work best or the material you want to print can involve quite a bit of trial and error. You also need to find out about things such as bed leveling, extrusion speed, fan settings, and nozzle purges. All of this gives you a lot of flexibility, but the learning curve can be a little steep. For printers of well-established brands on the higher end, the experience can be much more seamless, and you can often go from a 3D model to a physical part reliably in just a few clicks.

## Printing

When you transfer the G-code file to your printer and start the print, the printer will initially take a little time to heat up the nozzle and maybe the build plate. Depending on your settings, it will then print a purge line to prime the nozzle before printing the first layer of your model. The very first layer is the most critical of every print, so it is a good idea to carefully observe the printing of the first layer and try to spot any issues. If the material of the first layer does not adhere well to the build plate, or not enough or too much material comes out of the nozzle, you can stop the print, change parameters, and try again. Otherwise, the printer will keep printing, but without a good first layer, the entire print is poised to fail and just become a waste of time and filament. If the first layer is printed successfully, the chances of a successful print are high. The print can still fail due to a material jam, a clogged nozzle, or the printer running out of filament. So, it is good practice to check in with your printer periodically to avoid material waste due to a failing print.

If all goes well and the print finishes successfully, all you need to do is wait for the printer to cool down, take the part of the build plate, and remove any support material if necessary. *Figure 8.8* shows an example of a successfully printed part:

Figure 8.8 – The same cover, fresh off the 3D printer, printed from black PLA filament

For small parts, and if you have some CAD experience and a well-tuned 3D printer, the process from having an idea for a part to having a physical part can be as quick as 30 minutes.

## Additional tips for 3D printing

An important consideration when preparing 3D models for 3D printing is the **orientation** to print the part in. The mechanical properties of the final part are quite different in the two horizontal directions compared to the vertical direction. Since the printhead can move freely in the *x* and *y* directions, shapes in this plane can be printed with very high accuracy. However, the resolution in the vertical or *z* direction is limited by the layer height and is typically much lower. Further, the strength of the printed material itself is typically higher than the strength of the adhesion between the layers, which means that the printed part will be stronger in the *x* and *y* directions than it is in the *z* direction. Taking this inhomogeneity into account, you want to try to orient your part such that the features that need higher resolution as well as the main direction of physical stress on the part are parallel to the *x*-*y* plane.

Another factor that can influence the ideal part orientation on the build bed is **overhangs**. Because an FFF printer prints material layer by layer, the material of each layer needs to be laid on top of the previous layer. If the object you want to print has overhanging parts without material beneath to support them, these overhanging parts will likely fail to print. To mitigate this problem, you want to orient your part in a way that minimizes overhangs. You will not always be able to completely avoid overhanging parts. In this case, the slicer software can generate **support structures** to support the overhanging material. The support structure is usually printed with much less density than the actual part and is meant to break away once the print is finished. Some high-end printers have a dedicated nozzle and extruder for special support material that can be dissolved, allowing for more intricate support structures that might be hard or impossible to remove mechanically. Hobby-grade PLA printers, however, simply use the same nozzle and material to generate the support structures.

In the selection of printing parameters, you can trade off print speed versus print quality. During prototyping, it can be incredibly useful to print faster by increasing the printhead travel speed and layer thickness and decreasing the infill density. The infill is the amount of material that is printed inside the part. It is usually not necessary to print the part as a solid block, so to save on time, weight, and material, the slicer software can generate a criss-cross or honeycomb pattern with variable density, called **infill**. Lower-density infill will lead to a weaker part, but it can reduce the print time significantly. It can be a smart idea to initially print lower-quality versions of your part to allow you to iterate over your design faster, and only spend the time and material on a high-quality part, printed slowly and densely, when you are certain that you found the right design.

While PLA has many great properties that make it an excellent choice for many 3D prints, it is not very durable, especially when exposed to warm or hot temperatures. For parts that need to withstand high temperatures, you can consider using ABS filament. Most printers that can print PLA and have a heated build plate can also print ABS, but you might have to make some additional modifications, and you will need to make sure that the space in which your printer is located is well ventilated. ABS is better suited for parts that are subjected to repeated mechanical stress, such as the bumpers of an offroad robot, or parts that are exposed to high heat, such as a motor mount.

# Summary

In this chapter, we covered the basic principles of the mechanical design approach for DIY Arduino robots that maximize your chance of a successful project. These principles are similar to those that we have already talked about in the context of software development. In addition, we have looked at many important aspects of building the mechanical structure of your robot, namely ways to connect components, and materials to build custom components from. We have discussed, on a high level, the purpose and benefits of CAD and looked at the 3D printing toolchain in some detail. You can deepen your understanding of this by experimenting with various materials, glues, tools, and techniques to gain experience and intuition that will help you become an expert robot builder. You can take your robotic skills to the next level by getting familiar with CAD software and digitally modeling your first parts in 3D. If you have access to a 3D printer, you can learn a lot about the digital manufacturing process by experimenting with different printer settings and print parameters. These skills will allow you to eventually build more and more complex and unique robots quickly to bring all your robot ideas to life.

In the next chapter, we will switch gears from mechanical to electrical design considerations and discuss the power system of your robot, a crucial part of every robot design.

# Further reading

Following are some pointers to resources that can help you learn more about Arduino robot kits and 3D printing:

- A list with several examples of Arduino robot kits: `https://www.electronicshub.org/arduino-robot-kits/`
- The official Arduino Engineering kit, containing parts and instructions for interesting robotics projects: `https://www.arduino.cc/education/engineering-kit`
- A concise introduction video to 3D printing: `https://www.youtube.com/watch?v=biWEb8u1JYM`
- *Wikipedia* article on PLA material: `https://en.wikipedia.org/wiki/Polylactic_acid`

# 9
# Designing the Power System of Your Robot

Every electrically powered robot needs a battery and a power system for autonomous operation. This chapter will teach you how to select the right battery, how to safely integrate it into your robot, and how to generate different voltages for all the components of a robot. We will also cover how to safely handle high currents and integrate safety measures to prevent power-related damage to your robot. This chapter is structured as follows:

- Fundamentals of electrical power systems
- Understanding the power demands of your robot
- Safety considerations and hazards when working with high-power batteries
- Components and design of a robot power system

## Technical requirements

This chapter is full of information and will enable you to select the right components for the power system of your robot. However, the provided information is broadly applicable and does not only pertain to a specific example, and there is no specific hardware you need in order to get the most out of this chapter.

## Fundamentals of electrical power systems

Most DIY robots that you will build use electricity to power all their components. It is possible (but very unusual) to use other sources of energy for the actuators of your robot, such as compressed air, rocket motors, or combustion engines. However, for most DIY robotics projects, electric actuators are the most practical and cost-effective choice.

Since our robots run on electrical power, it is important to equip them with a reliable and appropriately designed power system to make them run. The power system's main job is to distribute the power from the batteries to the different components of your robot. The power system can be quite simple or quite complex, depending on the robot. It usually contains at least one voltage regulator or DC/DC converter to provide a stable logic voltage, but it might also contain sensors for monitoring voltages, currents, and power consumption, and safety mechanisms to prevent damage to the batteries or the robot. Because the power system is one of the core components of any robot, it needs to perform all its functions reliably and safely.

Even though it is a core subsystem of any robot, power system design is often overlooked by first-time robot builders. Nonetheless, it is important to have a good understanding of this part of your robot design. Problems with the power system can be hard to identify and therefore frustrating to deal with, so it is best to avoid them in the first place. Problems related to the power delivery to the various components of your robot can manifest with a wide range of symptoms, from excessive sensor noise to reduced motor torque. You might end up spending time and money replacing the sensors or the motors of your robot before discovering that the real problem was an insufficient voltage regulator.

Let us briefly cover the fundamentals of electrical power delivery to understand the requirements for our robot's power system.

## Voltage, current, and power

Voltage (*U*), electrical current (*I*), and electrical power (*P*) are the key figures to consider when talking about electrical power systems.

> **Note**
>
> In North America, the symbol *V* is commonly used for voltage rather than *U*. However, this easily leads to confusion between the voltage as a physical quantity and its unit of measurement, Volt (*V*). For that reason, we will use the symbol *U* for voltage, which is more commonly used in Europe.

The two key parameters that describe an electrical power source such as a battery, a phone charger, or a tabletop power supply are its **nominal voltage** and its **current capacity**. For example, your laptop charger might be labeled as having an output voltage of 20V and a current rating of 3.5A. That means that between its positive and negative output contacts, there is nominally a voltage of 20V. And your laptop can draw as much current from it as it needs, up to 3.5A. It is the load—the laptop, in this case—that determines how much current is flowing. But it is the source—the charger—that determines the voltage. The current rating on power sources such as power supplies and batteries is usually an upper limit at the nominal voltage. If your laptop, for example, tried to draw more current than the rated 3.5A from the charger, the charger will react by either dropping its output voltage, shutting itself off entirely, or—in the worst case—getting extremely hot and suffering thermal damage. That is just as bad for Arduino robots as it is for laptops.

We need to make sure that the current draw of our application (or robot) never exceeds the current capacity of its power source. The **electrical power** delivered by a power source, measured in Watts ($W$), is the product of voltage and current that is flowing: $P = U * I$. The maximum power that a laptop charger with a nominal voltage of 20V and a current capacity of 3.5A can deliver is, therefore, 70 W, for example. To put that into perspective, that amount of power is about 10% of horsepower. Components of the power system, especially batteries, are often characterized by their **power density**. The power density of a component describes how much power it can handle or deliver relative to its volume. Power density is often a function of cooling, since all electrical components tend to get hotter the more current flows through them, and what limits the power capacity of a component is often the temperature it can withstand. It is good to be mindful of the fact that any part of your power system, from the batteries to the voltage converters and even wires, can get hot. It is best to not enclose high-power components but to have them exposed to ambient air for a chance of passive cooling.

Any typical conductor, even a high-quality copper wire, has **electrical resistance**. The resistance ($R$) of a conductor describes how much voltage is needed to drive a certain amount of current through the conductor. This relationship is described in Ohm's law: $U = R * I$. The unit of resistance is *ohm*. A resistor of 1 ohm requires 1V of voltage to drive 1A of current through it. A resistor of 2 ohms requires 2V for the same current, and so on. If you know the resistance of a conductor and either the current or the voltage, you can compute the power that is lost and converted to heat along the conductor as $P = R * I^2$ (or equivalently, $P = \frac{U^2}{R}$). If both the resistance and the current are high, the dissipated power can be enough to heat up cables so much that things get damaged. This relation is why you often see thicker wires leading to a motor and thinner wires for sensors and electronics: thicker wires have less resistance and help keep the power dissipation low when currents are high. In contrast, if the current is low, such as for most sensors and other ICs, even a thin cable with more resistance will not lead to problematic power dissipation. Not only power is lost along a conductor, but according to Ohm's law, $U = R * I$ is also a voltage drop that depends on the resistance of the conductor and the current that is flowing through it. If you have a long, thin cable between the battery and the rest of your robot with a resistance of 1 ohm and your robot draws 5A of current, there will be a **voltage drop** of 5V along this cable. If your battery has a voltage of 12V, the robot on the other end of the cable will only see 7V.

> Tip
>
> The takeaway from all of this is: use thick wires wherever you expect large currents to prevent excessive losses, typically along the path from your robot's battery to the motors.

Cable manufacturers often provide useful charts that can give you an idea of how thick your cable should approximately be (see, for example, `https://www.jst.fr/doc/jst/pdf/current_rating.pdf`).

## Understanding the power demands of your robot

Most commonly, an Arduino robot requires electric power on at least two voltage levels:

- **Low voltage** (LV) or **logic voltage** of 5V or 3.3V (or both) for the electronic components
- **High voltage** (HV) or motor voltage for the electric motors—for example, 12V

Depending on the specific components of your robot, other intermediate voltage levels might also be required. Some sensors or analog circuits need an additional input voltage that is lower or higher than their logic voltage. Before you start designing the power system, you need to understand what voltage levels are required for your robot and how much current your robot will draw from them. As a rule of thumb, the current required at the low-voltage level is typically small, and the current drawn at the motor voltage level is orders of magnitude higher. For example, the ATMega microcontroller inside your Arduino Uno consumes around 0.1 W (or 100 milliwatts) of power, whereas an RC servo motor can easily consume 10 W or even more while moving or holding a static load. Because of that, it makes sense to select a battery with a nominal voltage that lies in the appropriate range for the motors of your robot (rather than the logic voltage). That way, you do not have to use a DC/DC converter to generate the motor voltage from the battery voltage. Given the high power demands of the motors, a suitable voltage converter might be large, costly, and require active cooling. In contrast, a DC/DC converter to generate the logic voltage from a higher battery voltage can be small and simple, since the logic components do not require much input current. This leads to a power distribution scheme as outlined in *Figure 9.1*.

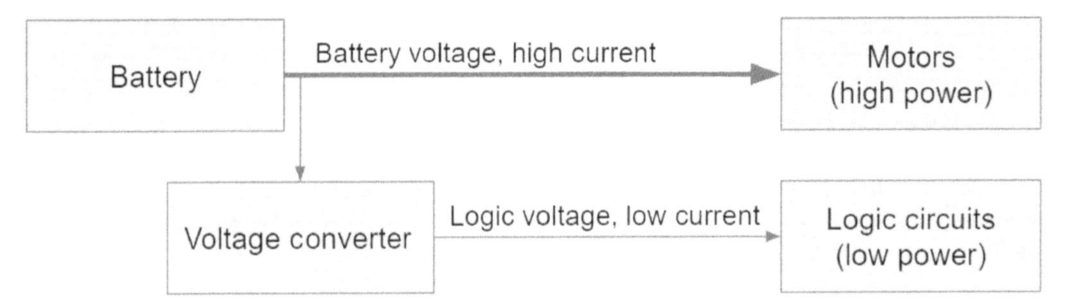


Figure 9.1 – The two main paths of electrical power in your robot

If you chose actuators for your robot that require different input voltages—for example, 12V motors for the chassis and 24V motors for the joints in a robotic arm—you can either use two different batteries with two different voltages or choose a battery voltage suitable for the actuators with the higher power demand. In this case, you can use a voltage converter to generate the input voltage for the actuators with the lower power consumption from the battery voltage. However, voltage converters that can provide enough current for motors are often large and expensive, making separate batteries for the different actuators the better choice in general.

> **Note**
>
> Do not forget to electrically tie the negative terminals of all batteries together to ensure that all components of your robot have a common ground. It is easy to miss this step, and it can lead to confusing problems that might take a while to track down.

# Safety considerations and hazards when working with high-power batteries

Since modern batteries can unleash a lot of power, the power distribution system is one of the components of a robot that can be dangerous for the robot itself, but also for the environment, and even for you. Let us spend a little bit of time looking at the risks that come from using electrical power and how we can mitigate them.

## Electric shock

If you touch two conductors and too much electrical current flows through your body, pain and lasting damage can occur. The best way to prevent this is to limit the voltages of your robot power system to safe levels. For DC power systems such as the ones in Arduino robots, the upper limit of what is considered safe to touch is 36V. If all the voltages in your robot are below 36V, it is safe to touch any part of your robot without the risk of serious eclectic shock (it might still tingle, especially when your fingers are not dry).

## Arcing

Even lower voltages can be dangerous, especially with power sources that have an extremely high current capacity—such as modern lithium batteries commonly used in robots. The current from a lithium battery can be so strong that it forms an extremely bright and hot blue arc if the two contacts of the battery are getting too close to one another. This bright arc can hurt your eyes when you look directly at it, and it can create enough heat to melt the metal of the conductor, adding the risk of molten metal flying around. One way this can happen is if you cut both the positive and the negative wire of a battery with the same cutter at the same time. Doing this might damage your eyes and the battery, and it will certainly damage the cutting tool. It is important to handle high-current power sources such as lithium batteries carefully.

> **Note**
>
> Always avoid accidental short circuits between the two contacts of a battery.

### Excessive temperature

When electrical current flows through anything (other than exotic superconductors), the electrical losses produce heat. The more current flows, the more heat is produced. If that heat is not dissipated at the same rate, the conductor's temperature increases. If a lot of current flows through a conductor, a wire, a transistor, or a motor, it can get dangerously hot, painful to touch, and may start to smoke or even set your robot on fire. To mitigate this risk, it is important to understand the current limitations of the components in your robot and use current limiting techniques such as fuses to make sure the safe level of current flow is never exceeded. What makes this a little more complicated is that it is usually no problem to have exceedingly high currents for a small amount of time. The drive motors of your robot, for example, might draw a lot of current for a moment during chassis acceleration, much more than during driving around. And that is not a problem, since it would take a while for dangerously high temperatures to build up. A practical and direct way of mitigating the risks from overtemperature is to monitor the temperature of critical components directly and shut off power when a threshold temperature is exceeded.

### Overvoltage

All electrical components have a nominal input voltage range. If the input voltage provided is lower than this range, the component usually simply does not function properly. But if the provided input voltage is too high, the component can get permanently damaged, might start smelling or smoking, or even catch fire. Most robots have a variety of components that need different input voltages, and thus your power system will have to provide different voltages. It is important to ensure that voltages that are too high for certain components are properly separated from these components to avoid damage. To avoid accidentally connecting a low-voltage component to a higher supply voltage, you can use color coding and different connectors for the different voltage levels in your robot. You can also use appropriately sized Zener diodes wired in parallel to a sensitive component as transient overvoltage protection, similar to how a fuse wired in series can protect against overcurrent. A crowbar circuit (`https://en.wikipedia.org/wiki/Crowbar_(circuit)`) or dedicated ICs such as *Onsemi*'s NCP346 (`https://www.onsemi.com/pdf/datasheet/ncp346-d.pdf`) can provide reliable overvoltage protection in more challenging circumstances.

## Components and design of a robot power system

Now that we have a good understanding of the various considerations that go into designing a power system, let us look more closely at the individual components that may go into it. We will organize the components into three categories: the core components that are used to *distribute* the power, components that can be used to *monitor* the power system's status, and components to *control* the power system and ensure its safe operation.

## Power distribution

Distributing the power from the batteries to the components that need it is the fundamental function of the power systems. Everything else is optional.

### *Batteries*

The batteries are what power your robot. The two most important parameters of a battery are its **nominal voltage** and its **capacity**.

During normal operation, the voltage of a battery will vary: it will be a little higher than nominal when it is fully charged, and lower when it is empty. It also changes depending on the current that is drawn from the battery. Due to the battery's internal resistance, the voltage will drop lower the more current is being drawn. The exact magnitude of these effects varies from battery to battery, but it is important to keep in mind that the *battery voltage is not constant*. If you build a balancing robot—for example—that uses a feedback control system to control its motors, you might spend a lot of time tuning the controller parameters only to find out that it works fine when the batteries are full, but performance is much worse when the batteries are drained. This might be because the controller performance depends on the battery voltage, and the voltage is constantly changing.

The capacity of a battery describes how much energy it can store when it is fully charged. The battery capacity is commonly described in **milliamp hours (mAh)**. This quantity tells you how much current (in mA) the battery can supply for 1 hour. For example, a battery with a capacity of 2,000 mAh can sustain a discharge current of 2A for 1 hour. After that, the battery is empty. At twice the current (4A), the battery would only last half an hour. To compute the amount of energy that the battery can store, you need to multiply the capacity by the battery's nominal voltage. If the battery in the previous example had a nominal voltage of 12V, its maximum energy content would be 24 **watt-hours (Wh)**. If you know the power consumption of your robot, you can pick a battery with the capacity that gives you the required runtime for your application.

Other important characteristics of batteries can help you choose the right one for your robot. Current capacity is certainly one of them. The current capacity describes how much current the battery can continuously supply (until it is empty), without overheating or drastically shortening its lifespan. This rating is usually expressed as a **C rating**. A discharge rating of 10C, for example, means that the battery can continuously supply 10 times the current that it would take to discharge it in 1 hour. With the battery of our running example that has a capacity of 2,000 mAh, a current rating of 10C means that it can continuously deliver 20A. This amount of current would deplete it in one-tenth of an hour, or six minutes. If your robot tries to draw more current, the battery will either get hot or the voltage will drop by a lot, or both. The lifetime of the battery might also get reduced as a result. Modern lithium batteries can achieve very high output current ratings. It is common to find a battery with a capacity on the order of 5,000 mAh and a 50C rating on a DIY robot, which means that it is capable of 250A of continuous current. That is an extremely large current, enough to weld steel. In contrast, the maximum current of household non-rechargeable batteries such as the common AA or AAA types is two orders of magnitude lower. Even during a short circuit, these batteries only supply a few amps of current. This makes them very safe to use, but a poor choice for powering the motors of DIY Arduino robots:

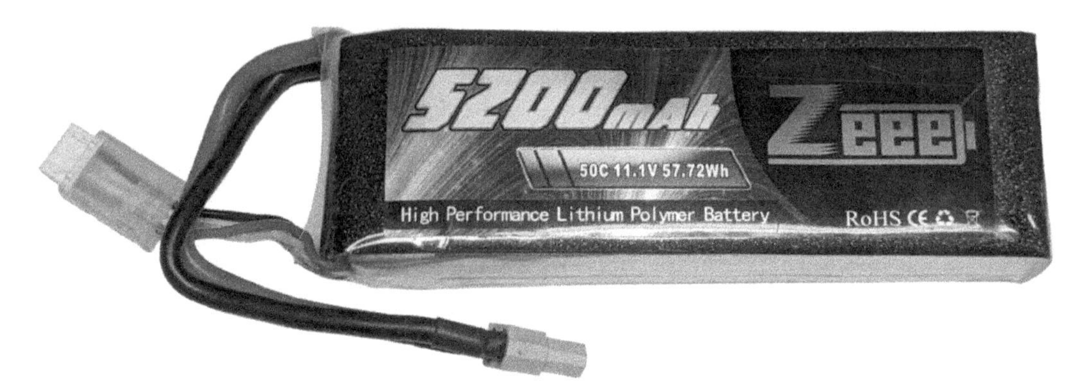


Figure 9.2 – A three-cell lithium polymer (LiPo) battery with a nominal voltage of 11.1V, a capacity of 5,200 mAh, and an impressive continuous current rating of 50C or 260A

There are many types of rechargeable batteries that can be used for mobile robots. We can categorize them as **lead-based**, **nickel-based**, and **lithium-based**. Lead-based batteries are very heavy, cheap, and easy to use. They are typically used when size and weight are not important, such as in robotic boats. Nickel-based batteries have a higher power density and are a little lighter. However, they have mostly been replaced by lithium-based batteries.

> **Note**
>
> Lithium-based batteries in the form of **lithium-ion** (**li-ion**) batteries in cylindrical cells or LiPo batteries in pouch cells have become the dominant battery type for DIY robots.

Lithium batteries require more careful handling and dedicated chargers capable of cell balancing (keeping the voltages equal across cells), but their power and energy density are unmatched by any other technology. Thanks to mass production driven by their ubiquitous use in consumer electronics and electric vehicles, these batteries are also very affordable.

### *DC/DC converters*

The batteries are the energy source of your robot. However, their voltage varies depending on their state of charge and operating conditions, and it might not match the input voltage range of some components of your robot. For example, you cannot directly power your Arduino Uno board from a 12V battery. To generate a stable logic voltage of 5V for the electronic component of your robot, you need a voltage converter—more precisely, a converter that takes in a DC voltage within a certain range and outputs a tightly controlled output voltage. Voltage converters, sometimes called switching voltage converters or DC/DCs for short, are different from linear voltage regulators. Linear voltage regulators such as the one on the Arduino Uno board can also generate a lower voltage from a higher input voltage, but they are very inefficient at doing so. This means that they produce a lot of heat and are not very power dense. Switching voltage converters, in contrast, are extremely efficient, usually do not require much by way of cooling, and can handle much more power than linear regulators of the same size.

To find the right converter for your application, you need to consider the input voltage range, the output voltage, and the output current capacity. The input voltage range should be larger than the voltage range of the battery you are using. If you use an 11.1V LiPo battery with a voltage range from 9.5V to 12.5V, a suitable voltage converter should cover at least an input range from 9V to 13V. Modern DC/DC converters are extremely good at tightly regulating their output voltage. A converter with a 5V output voltage, for example, will always produce a voltage close to 5V, no matter where in the input voltage range the actual input voltage lies. Voltage converters are specified for a certain maximum output current, and you want to find one that can supply enough current for your application. *Figure 9.3* shows a DC/DC module capable of 3A continuous output current:

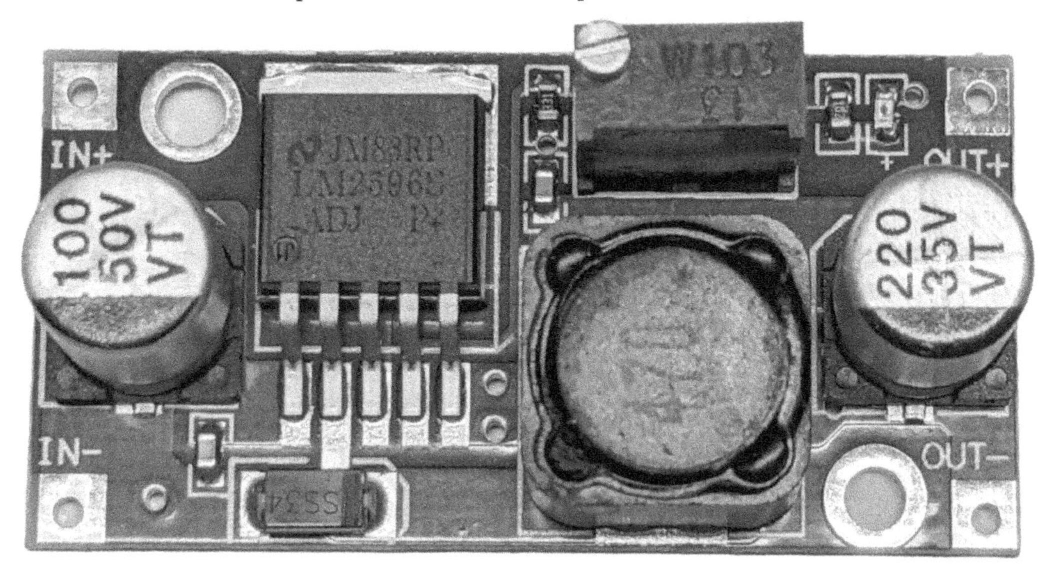


Figure 9.3 – A low-cost DC/DC module with an adjustable output voltage that can provide up to 3A of output current

At a minimum, a voltage converter has three terminals: ground, input voltage, and output voltage. If the input and output voltage share the same ground, the converter is a **non-isolated** converter. In some applications, it is desirable to achieve galvanic isolation between the input and the output voltage levels. This can be done with an **isolated** voltage converter. An isolated converter has four terminals: input voltage and input ground, as well as output voltage and output ground. Some DC/DC converters have additional pins for monitoring and control. It is common for these types of converters to have an **enable input** that can be used to turn the output voltage on and off, like the one shown in *Figure 9.4*.

Figure 9.4 – A small, breadboard-mounted DC/DC module with fixed output voltage and an active low shutdown input

Using the enable or shutdown input, your Arduino can control when certain voltages are provided, which can be useful for system diagnosis, safety, or to reduce the power consumption of your robot under certain conditions.

### *Switches*

It can be extremely useful to have manual switches that allow you to turn your entire robot, or certain subsystems, on or off. However, switches that can handle the amount of current that a mobile robot requires can be quite large, which is why it is uncommon to see a master power switch on a mobile robot that sits in line with the main system current. However, it is common to see switches as an input element to control the power system. The Arduino brain of your robot can read the state of the switches and enable or disable parts of the power system accordingly, using the enable inputs of DC/DC converters or **Metal Oxide Semiconductor Field Effect** (**MOSFET**) transistors. You can also use manual switches to control the enable pins of DC/DC converters directly.

## Monitoring

It is often useful for your Arduino program to know at least a little bit about the state of the power system. If it knows the battery voltage, it can estimate the battery's state of charge and warn you when the battery is running low. If it knows the temperature of certain components, it can reduce the motor power to prevent overheating when necessary. It can also transmit power telemetry to a base station if you control your robot remotely, or you can log it to an SD card for later examination. The three quantities that are most useful for monitoring the power system are voltages, currents, and temperatures. Let us look at how your Arduino can sense all three of them.

### Voltage sensing

Sensing voltages is the easiest of the three since every Arduino has a built-in voltage sensor in the form of an **analog-to-digital** (**ADC**) converter. All we need to do to measure a voltage—for example, the battery voltage—is to map that voltage to the input voltage range of the ADC with a **voltage divider** and convert it back to the original voltage range in our program. A voltage divider consists of two resistors in series between the voltage we want to measure and ground. These resistors should be in the 10k ohm range or higher to make sure the current through them is low. All that matters is the ratio between the two resistors, not their absolute values. You can size the resistors of the voltage divider such that the voltage across the low-side resistor (the one connected to GND on one side) will never exceed the maximum input voltage of your Arduino's ADC, and then use the ADC to measure this voltage like in the setup depicted in *Figure 9.5*.

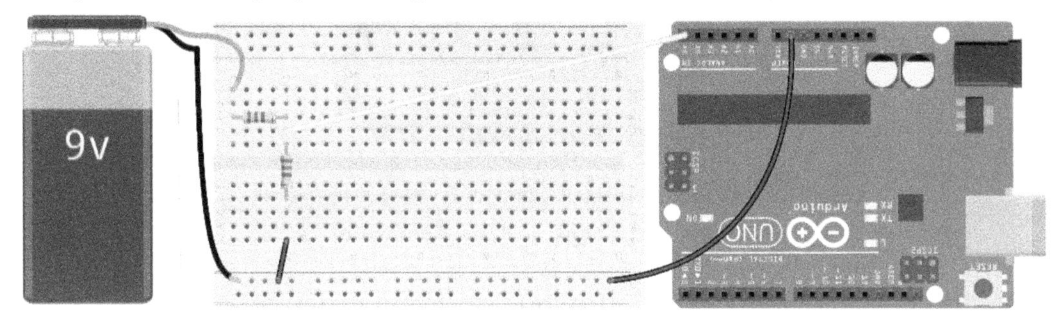


Figure 9.5 – Measuring the voltage of a 9V battery with an Arduino Uno and a voltage divider

Knowing the ratio between the resistors and the resolution of your ADC, you can convert the ADC reading back to the voltage that you want to measure. We will go over the resistor selection and the required code to work out the voltage in our example project in *Chapter 12, Building an Advanced Line-Following Robot Using a Camera*.

### Current sensing

Sensing a current is more difficult than sensing voltages. The main problem with current sensing is that your sensor needs to be *in series* with the current you want to measure, meaning all that current needs to flow through your sensor. The easiest way to measure currents is to use a resistor and the Arduino's built-in ADC as a current sensor. The sense resistor (sometimes ambiguously called shunt resistor) should be placed where the current path connects back to the ground, such that one end of the resistor is connected to the ground. The current flow through the resistor will cause a voltage across the resistor according to Ohm's law, and you can use the ADC of your Arduino and convert the voltage back to a current reading. The circuit is the same as the low-side part of the voltage divider shown in *Figure 9.5*, just with a much lower resistance value. However, the voltage signal will be quite small, and the low-side sense resistor effectively changes the GND potential for the upstream components, which can be problematic. We can avoid these problems by using dedicated high-side current monitors with built-in analog amplifiers such as *Texas Instruments*' INA169, shown in *Figure 9.6*:

Figure 9.6 – Adafruit's breakout board for TI's INA169 high-side current monitor, a handy analog current sensor

An even more elegant alternative are Hall-effect-based current sensors that do not require a sense resistor and can be placed anywhere along the current path. Infineon's ACS723, for which SparkFun makes a handy breakout board, is an example of this type of sensor (`https://www.sparkfun.com/products/13679`).

### *Temperature sensing*

Temperature sensors can be crucial to protect components such as batteries, motors, and DC/DC converters from thermal damage. When your Arduino program detects critical temperatures in any of these components, it can halt the robot's operation and temporarily shut off the hot components to prevent overheating and damage. You can purchase small temperature sensor boards with digital interfaces such as I2C, or standalone thermistor-based analog temperature sensors such as Microchip's MCP9700. The latter is often easier to attach directly to the component that you want to monitor. Placing the sensor as close to and in direct contact with the point where you want to measure the temperature is critical to get an accurate temperature measurement.

## Control and safety

As we discussed earlier in this chapter, depending on the type of batteries your robot uses, an improperly designed or malfunctioning power system can cause a lot of damage, all the way up to setting your robot on fire. Some accidents can be prevented by letting your Arduino not only monitor but also control the power system. And if all else fails, your robot's power system hopefully includes passive safety elements such as fuses.

### *MOSFETs*

The two main ways your Arduino can control the power system are by switching DC/DC converters on and off using their enable inputs, and by switching other subcomponents or branches of the power system on or off using power MOSFETs. MOSFET refers to a class of transistors that are easy to use and have exceptionally low internal resistance and high current-carrying capacity. MOSFETs can be either the N-channel or the P-channel type. You can place these transistors in the path of the electrical current you want to control, and they will block the current flow by default. Only when your Arduino turns them on can the current flow. MOSFETs have three pins: *gate*, *source*, and *drain*. If your Arduino supplies a voltage between the gate and source, a large current can flow from the drain to the source. You can place the MOSFET between the supply voltage and the load, or between the load and the ground. The former is called a high-side switch, and the latter a low-side switch. Low-side switches are easier to implement since your Arduino can drive the gate of a low-side N-channel MOSFET directly:

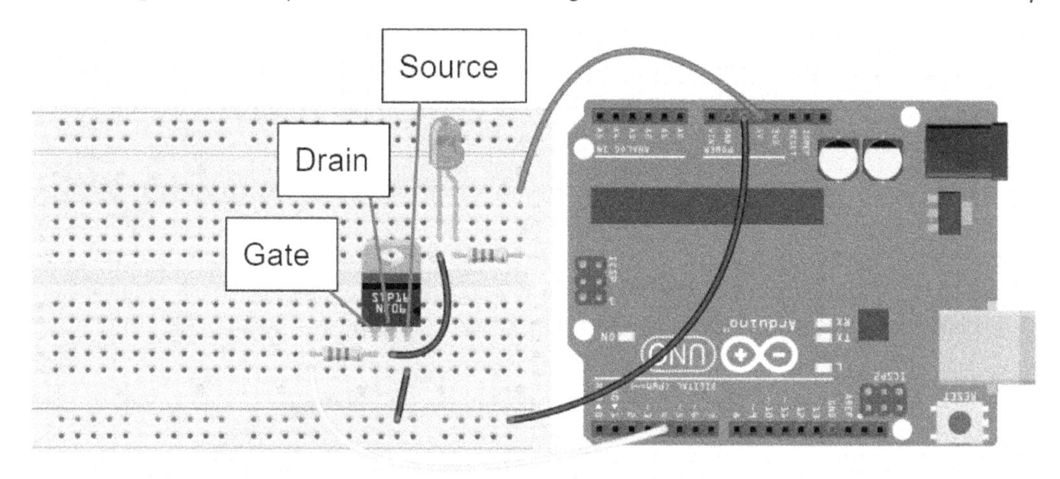


Figure 9.7 – An Arduino controlling current through an LED using a low-side N-channel MOSFET as a switch

For a high-side transistor, some added circuitry or a dedicated high-side gate driver might be required, depending on the voltages involved. Electromechanical relays can be an alternative to MOSFETs, but they are usually much larger, more expensive, and require dedicated driver circuitry to drive their inductive inputs, just like the ones we use for DC motors.

### *Fuses*

To reliably prevent too much current from flowing through a current path in the power system of your robot, you can use fuses. Electrical fuses are passive components that conduct current with little resistance under normal conditions. However, if too much current flows, they heat up and automatically interrupt the current flow. Fuses can be either single-use or resettable. A single-use (or non-resettable) fuse gets permanently destroyed when too much current flows, and permanently and completely interrupts the current path. In contrast, a resettable fuse drastically increases its resistance when the current exceeds a certain limit but does not completely interrupt the current flow. If the current source is removed and the fuse can cool down, it automatically resets itself to low-resistance. Fuses are primarily characterized by the maximum current they can carry before tripping, called the trip current. For example, a 2A fuse will trip and interrupt the current flow once it exceeds 2A. A secondary characteristic is how fast the fuse trips. Fast-acting fuses will trip as soon as the overcurrent occurs. If your application needs to allow for short overcurrent events (for example, to accommodate inrush currents when initially connecting the battery or when starting motors), a slow-acting fuse can be the better choice. There are also **electronic fuses** or **eFuses** (`https://www.st.com/en/power-management/e-fuses.html`), ICs that integrate current sensing, power switching, and a digital interface to communicate with a microcontroller. These are more complicated to use but have several advantages over passive fuses. For example, they can often be used as automatic, highly efficient inrush current limiters, and to enhance your power system's monitoring capabilities. *Infineon* makes an Arduino shield that lets you integrate its BTS7002-1EPP IC into your Arduino robot easily (`https://www.infineon.com/cms/en/product/evaluation-boards/shield_bts7002-1epp/`).

Fuses come in a wide range of shapes and footprints, covering **Surface-Mount Technology** (**SMT**), **Through-Hole Technology** (**THT**), and socket-mounted form factors, among many others. Cylindrical cartridge fuses, either made from a single wire housed in a glass tube or from a ceramic material, are widely available in various current ratings and are easy to integrate into your robot. You can simply make a fused adapter cable between the battery and the robot using an inline fuse holder (see *Figure 9.8*).

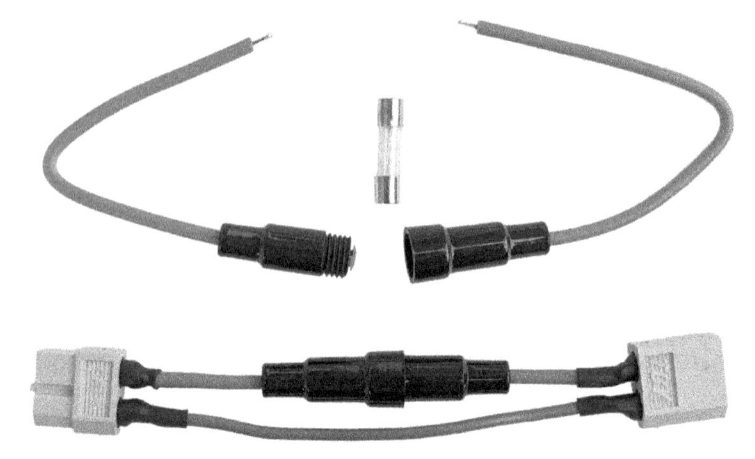

Figure 9.8 – Top to bottom: A 5A glass tube cartridge fuse, an inline fuse holder, and a fused power cable to safely connect a robot to a high-power Lithium battery

Fuses are often used as the last line of defense against damage caused by too much current. You should place a fuse in series with the main battery of your robot, with a trip current slightly larger than what you would expect your robot to ever draw. If due to malfunction or mechanical damage a short circuit between the battery and the ground occurs, the fuse will trip and cut the battery power to prevent damage. Without a fuse, the excessive current could easily cause a lot of damage, up to setting the entire robot on fire. Given that fuses are cheap and are a great way to prevent catastrophic damage, it is always good to use them in your robot's power system.

## Putting it all together

To put everything that we learned in this chapter into context, let us look at *Figure 9.9*, which shows an overview of an exemplary power system. This robot uses two types of motors, one that runs on 12V and one that runs on 24V. To power these two, it has two sets of batteries: a 12V battery and a 24V battery. Each battery is protected against overcurrent by a passive fuse directly after the battery in the current path, and the Arduino can sense the battery currents with Hall-effect current sensors. The batteries feed the motors via their respective motor controllers. Each of these branches can be switched off or on via a low-side MOSFET switch that is controlled by the Arduino. This allows your robot to save energy when the motors are not needed, and to only enable them when it is safe to do so. The 12V battery feeds a 5V DC/DC converter that is always enabled to power the Arduino board. It also powers additional DC/DC converters with enable inputs that are controlled by the Arduino. The Arduino only turns them on when they are needed. These converters feed all the additional 5V and 3.3V low-power electronics. What is not shown in this diagram are voltage dividers that map the battery voltages to the range from 0-5V so that the Arduino's ADC can measure them. This allows the Arduino to shut down the motors when the battery voltages drop below the lower limit of the nominal voltage ranges. The arrows in the high-current paths from the batteries to the motors are heavier to indicate that thicker wires should be used here than in the logic-power section of the system. This is to minimize losses generated by the high current.

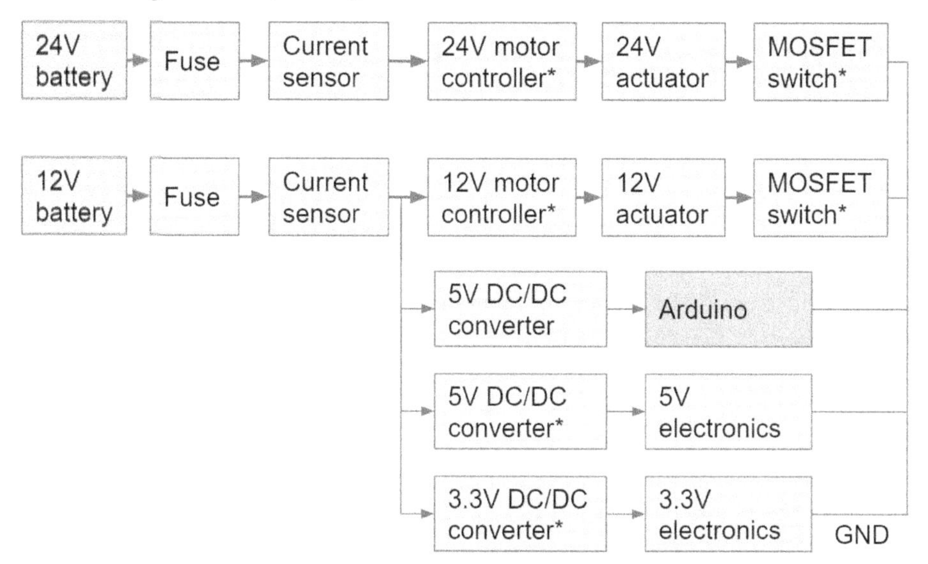


Figure 9.9 – Schematic overview of an exemplary robot power system; components with an asterisk (*) are used by the Arduino to control the power system

This example shows quite an elaborate power system. Not every Arduino robot needs a power system this complex. But now that you know the core components and the reasoning behind their usage and selection, you can pick the aspects and components that make sense for your robot project.

## Summary

In this chapter, we discussed the importance of careful power system design for your Arduino robots, especially when using powerful rechargeable lithium batteries that have the potential to cause substantial damage by virtue of their immense power density and current capacity. We got to know the key components of a robot's power system that can serve you well even in advanced and power-hungry robots. Not all Arduino robots need an elaborate power system, and the smallest ones can even run off USB power. But whenever you use powerful batteries in your robot designs, it is a good idea to spend some thought on a power system that makes your robot safe and reliable.

In the next chapter, we will learn how to give your robot the ability to express itself with LEDs, displays, and sound.

## Further reading

There is much more that we could discuss about a robot's power system and electrical power in general. The following resources can help you get started to dive deeper into those topics:

- A blog post about different types of voltage regulators: `https://www.digikey.com/en/maker/blogs/2020/what-is-a-voltage-regulator`
- An article on Ohm's law, a fundamental relation in electronics, on Wikipedia: `https://en.wikipedia.org/wiki/Ohm%27s_law`
- A comprehensive list of battery types with links to their specific articles on Wikipedia: `https://en.wikipedia.org/wiki/List_of_battery_types`
- A brief safety guide for working with different types of batteries: `https://iaeimagazine.org/electrical-fundamentals/a-safety-guide-for-working-with-batteries/`
- Lessons learned from the **Jet Propulsion Laboratory** (**JPL**) robot battery fire in 2015: `https://llis.nasa.gov/lesson/23701`

# 10
# Working with Displays, LEDs, and Sound

To make robots more interactive and also help with system debugging, it is often helpful to let them display their status with **light-emitting diodes** (**LEDs**), displays, or audio output. This chapter will show you how to use different types of displays, LEDs, and speakers to build interactive robots and is structured as follows:

- Advanced use of LEDs
- Using different types of displays
- Adding sound output to your robot

We will see hardware and code examples that demonstrate all the techniques we cover in this chapter.

## Technical requirements

There are several examples in this chapter that show how to use different hardware. Besides your Arduino Uno and your PC, you will need at least a solderless breadboard, an LED, a 220-Ohm resistor, a potentiometer, and a few jumper wires. To follow the more advanced examples, you will also need either three NPN bipolar transistors or three N-channel MOSFETs, a multicolor (RGB) LED strip, and a character **Liquid-Crystal Display** (**LCD**). For the sound examples, all you need is a piezo buzzer.

## Advanced use of LEDs

LEDs are a wonderful way of adding expressiveness to your robot. They are a cheap, easy to control, and very power-efficient way to create light in a large variety of colors. LEDs can be incredibly useful both as status indicators of your robot and as interactive elements that let your robot express its intention or its mood.

We have already learned how to control the LED of your Arduino Uno that is connected to pin 13, and how to let it blink using our custom `Blinker` class that complies with the cooperative multitasking paradigm. But there is a lot more that we can do with LEDs, and a lot more to learn on how to control them and integrate them into your robot. So, let us get started!

## Dimming LEDs

In our previous examples, all we did with the LED was turning it either on or off. This is usually enough to convey a status with an LED, or signal that the program is running. But there are many more effects we can create by learning how to change the brightness of an LED.

Fundamentally, all we can do with a digital output pin is to turn an LED either on or off; we cannot make it light up with variable brightness. But we can combine the capability to switch it on and off very rapidly with the **Persistence-of-Vision** (**PoV**) effect that the human eye is susceptible to: our eyes have limited temporal resolution, which means that we cannot see changes in an LED (or anything) that happen too fast. If an LED, for example, blinked with a very high frequency, we cannot tell that it is blinking. Instead, we will perceive the LED as persistently turned on, but with a lower brightness than if it was constantly lit. The relative brightness is controlled by the ratio of the ON and the OFF duration within one blinking cycle, often called the duty cycle. A duty cycle of 100% means that the LED is constantly on, 50% means that the ON and OFF durations are equal, and a 0% duty cycle means that the LED is off. This is exactly what PWM does, and we have used this before to control the power of a DC motor. The downside of using hardware PWM is that we can only use it on pins that are PWM capable and that we cannot easily select the PWM frequency (more on that aspect in *Chapter 13, Building a Self-Balancing, Radio-Controlled Telepresence Robot*). So, let us use this opportunity to take what we have learned so far and write the code that allows us to generate *software PWM* on any digital output pin. This will enable us to dim the LED that is built into your Arduino board and connected to pin 13, even though pin 13 is not a PWM pin. We do not even have to write a lot of code; we can instead create a new `Dimmer` class as a slightly modified version of our `Blinker` class. The two main modifications we need to make to the `Blinker` class are the following:

- We need to be able to control the duty cycle. So far, we only varied the blinking frequency with a constant duty cycle of 50%. So, we will add a new `duty_cycle_` member variable to our class and a `set_duty_cycle()` function that lets us control it.
- We need to blink much faster than before, so instead of basing the blinking state machine on the `millis()` function for millisecond resolution, we will base it on `micros()`, which gives us microsecond resolution.

There are, as usual, many ways we could implement this. The trade-offs are primarily code readability versus efficiency. Because our Arduino will need to blink the LED at a very high frequency, it is important to implement these new capabilities efficiently. Therefore, we will avoid the use of floating-point arithmetic since this requires a lot more CPU cycles than integer arithmetic. The key to the `Dimmer` library implementation is that we fix the blink interval to `1024` microseconds, which is fast enough to exploit the PoV effect. This change affects the class constructor, which now looks like this:

```
Dimmer::Dimmer(int led_pin, int duty_cycle) {
  led_pin_ = led_pin;
  blink_interval_ = 1024;  // microseconds
  duty_cycle_ = duty_cycle;
}
```

To control the duty cycle, we update the `last_blink_time_` member variable inside of the blink state machine, depending on whether we are in the `LED_ON` or `LED_OFF` state and based on the newly added `duty_cycle_` member variable. The modified `blink_task()` function for the `Dimmer` class looks like this:

```
void Dimmer::blink_task() {
  switch (led_state_) {
    case LED_OFF:
      if (duty_cycle_ > 0) {
        digitalWrite(led_pin_, HIGH);
      }
      led_state_ = LED_ON;
      // Will be switched off after duty cycle has elapsed.
      last_blink_time_ += duty_cycle_;
      break;
    case LED_ON:
      digitalWrite(led_pin_, LOW);
      led_state_ = LED_OFF;
      last_blink_time_ += blink_interval_ - duty_cycle_;
      break;
  }
}
```

We also add a new `set_duty_cycle()` function that lets us change the private `duty_cycle_` variable. With these changes, we can now set the duty cycle of the fast blinking between `0` (0%) and 1023 (100%) to control the brightness of an LED. This function is a simple setter that constrains the duty cycle to the range of valid values:

```
void Dimmer::set_duty_cycle(int duty_cycle) {
  duty_cycle_ = constrain(duty_cycle, 0, 1023);
}
```

As an easy way to demonstrate this new capability, we can instantiate a `Dimmer` object and, in the `loop()` function, set its duty cycle to the the current time in milliseconds modulo `1024`. This will generate a sawtooth profile, going from `0` to `1023` (or 0% to 100% duty cycle) at a little more than 1 Hz:

```
#include "Dimmer.h"

//Instantiate a dimmer on pin 13.
Dimmer my_dimmer(13, 500);

void setup() {
  my_dimmer.begin();
}

void loop() {
  // Generate saw-tooth brightness profile.
  int duty_cycle = millis() % 1024;
  my_dimmer.set_duty_cycle(duty_cycle);
  my_dimmer.blink();
}
```

You can additionally connect a potentiometer and set the duty cycle of the LED to the analog reading of the potentiometer input instead. This gives you a great way to experiment with the perceived brightness. To this end, simply replace the first line in `loop()` with the following line:

```
int duty_cycle = analogRead(A0);
```

You will find that your eyes are much more sensitive to changes in brightness when the LED is darker (in the lower range of duty cycles) than when it is brighter.

## LED circuits

So far, we have only ever used one LED—the one that is built into the Arduino Uno board, connected to pin 13. If you build your own robot and want to display its status or maybe its expressions with LEDs, you will want to add external LEDs that you can place where you need them. Let us look at how we can control external LEDs.

### *Standard LED*

Connecting an external **standard LED** to any pin of your Arduino so that you can control it with the `digitalWrite()` function is easy: one leg of the LED needs to be connected to the Arduino's pin, and the other to either 5V or GND. It is very important though to add a *current-limiting resistor* in series with the LED; otherwise, it might easily get damaged. *Figure 10.1* shows how to connect an external LED to your Arduino Uno:

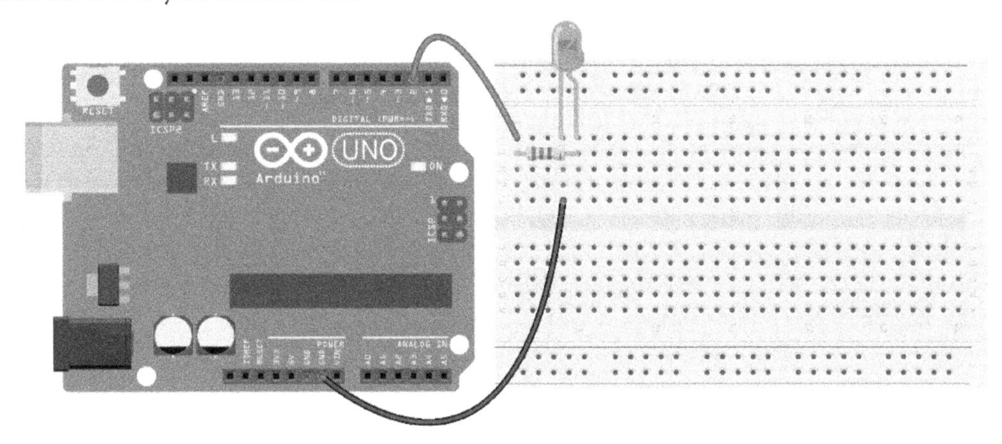


Figure 10.1 – Connecting a standard LED to an Arduino output pin via a current limiting resistor

To determine the correct value for this resistor, you need to know the **forward voltage** of the diode as well as the operating current. You can find the forward voltage and the maximum operating current in the LED's datasheet. To calculate the resistor value *R*, we can use **Ohm's law** in the form of $R = \frac{U}{I}$, where *U* is the difference between the supply voltage and the forward voltage, and *I* is the desired operating current that should not be higher than the maximum operating current from the datasheet.

**Example**: Let us assume the datasheet of our LED states a forward voltage of 2V and a maximum current of 20 mA, and we want to drive it from the output pin of a 5V Arduino. To give the LED some margin, we are targeting an operating current of 15 mA. In this case, the ideal resistor value is:

$$R = \frac{5V - 2V}{0.015A} = 200\,Ohm$$

The next closest standard resistor value is 220 Ohm, so we would use that. You could use a larger resistor to reduce the current accordingly, which would result in less brightness. Using a resistor to control and limit the LED current is an easy and highly practical way of doing it. However, the current and associated voltage drop across this resistor creates heat and losses, and if you want to drive more powerful LEDs, these losses might not be tolerable. In that case, you might want to use a dedicated LED driver chip or a discrete circuit that directly controls the current with power MOSFETs. This approach can be significantly more efficient than using a series resistor. However, for low-power LEDs, a series resistor is a practical and by far the easiest solution.

### *High-power LED*

The brighter LEDs become, the more current they require. In our previous example, we drove an LED with 15 mA, which is a very small amount of current. However, the Arduino Uno can only continuously provide 20 mA of current on a digital output pin, so we cannot drive much more powerful LEDs with this simple method. To drive very bright LEDs that might require currents of several 100 mA, we need a transistor to handle that current, just like we need a motor driver to drive a DC motor. Fortunately, driving an LED is simpler than driving a DC motor, since current only needs to flow in one direction and LEDs do not have the significant inductance that DC motors have.

The simplest way to drive a high-power LED is by using a transistor—namely, an N-channel MOSFET or an NPN bipolar transistor—as a *low-side switch* to switch the LED on or off. In that case, all the Arduino needs to do is drive the **gate** (for a MOSFET) or the **base** (for a bipolar transistor) of the transistor. The transistor handles the high LED current. This requires very little current from the Arduino, and you can select a transistor that matches the current requirements of your LED. The following diagram shows this circuit:

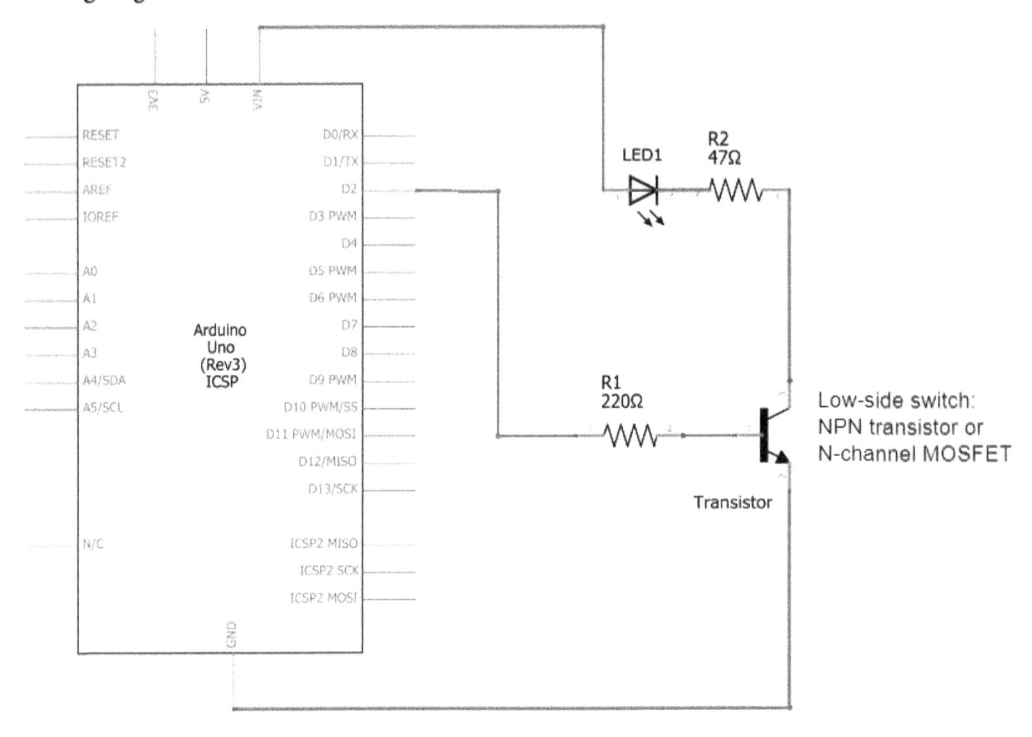


Figure 10.2 – Schematic of a low-side switch implemented with an NPN transistor to control the current through an LED

The schematic makes it clear why the transistor is a low-side switch in this circuit: in the current path from VIN to GND, the transistor is on the GND side—the low side. The LED current comes from the VIN pin of the Arduino, goes through the LED and a current limiting resistor, and then through the transistor's collector-emitter path to GND. The Arduino only needs to drive a very small current through the transistor's base-emitter path to switch the LED current on.

The following is the same circuit, assembled on a solderless breadboard:

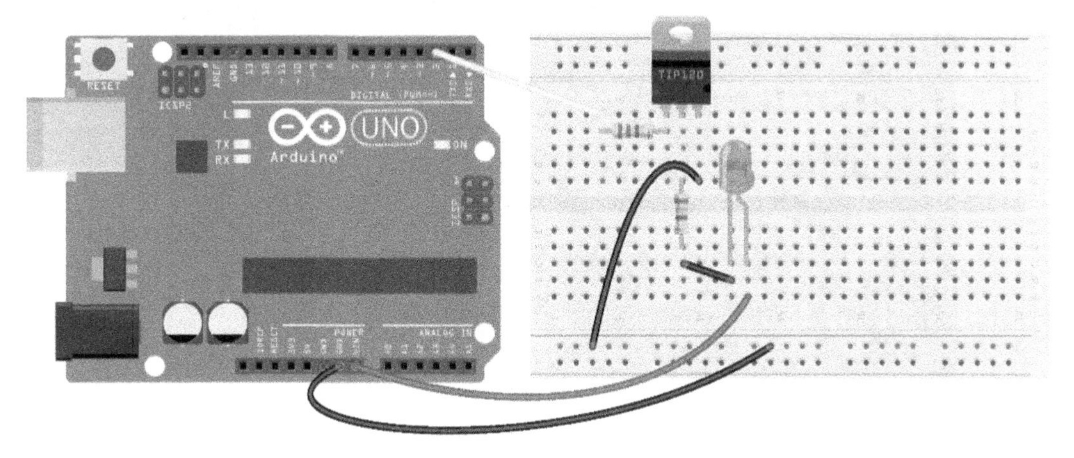


Figure 10.3 – Controlling a high-power LED with a transistor acting as a low-side switch

The Arduino program is the same whether you drive an LED directly or if you use the Arduino to drive the transistor. One thing to keep in mind with this setup is the power that is dissipated by the LED's current limiting resistor (*R2*) and transformed into heat. You can compute this power dissipation *P* based on the current *I* that flows through the resistor and the value *R* of the resistor as $P = R * I^2$. For example, if you have a 100 Ohm resistor and an LED current of 50 mA, the resistor's power dissipation will be 0.25 W, and you will need to use a resistor that is rated for that much power to not risk overheating. Because the power is a function of the square of the current, it quickly becomes quite large for more powerful LEDs that require more current, and resistors that can continuously dissipate a lot of power are costly and large. They also get quite hot, to the point where it can be dangerous to touch them. For those reasons, it is better to use dedicated constant-current LED driver circuits or ICs to drive high-power LEDs, such as the ones you would use as headlights on a robot that operates outdoors in the dark.

## Multicolor LEDs

Other than the brightness, multicolor LEDs let you control the color of the light they emit. Fundamentally, a single LED can only produce light of one color, so to produce a multicolor LED, multiple LEDs are combined in one chip or one small assembly. The most common combination consists of a red, a green, and a blue (RGB) LED. By mixing the light of these three colors, which means turning on the three LEDs at different brightness levels, we can create any color of light that we like. The following picture shows a close up of an RGB LED display, rendering different colors:

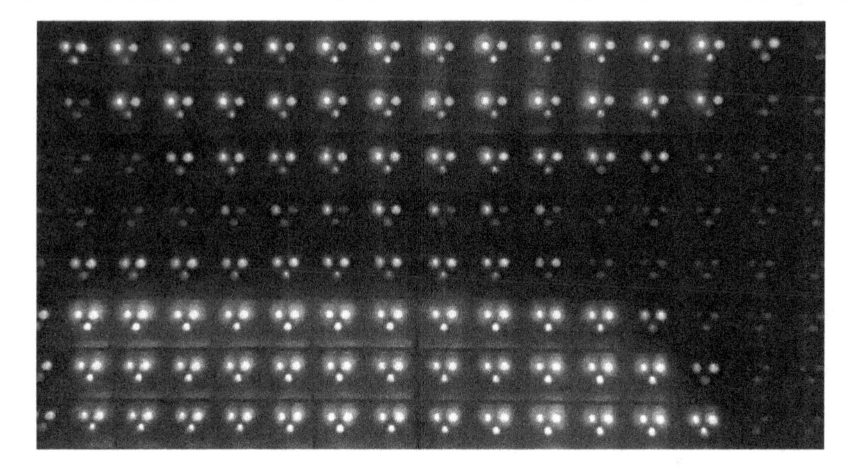

Figure 10.4 – A close-up image of an LED color display; color and brightness are varied by controlling the RGB LEDs of each pixel individually

This is how most screens produce colors, too. If you look very closely (with a magnifier or microscope) at the screen of a TV, a computer, or a phone, you will see that each pixel of the screen consists of multiple subpixels that can produce either red, green, or blue light. Controlling the relative brightness of these subpixels (for example, by blinking them with varying duty cycles) is what allows the pixel to light up in any color.

### Multicolor LED example

You can create a multicolor LED module by simply combining three LEDS, a red, green, and blue one, closely together and driving them individually with our `Dimmer` class. Or you can buy an integrated module that already combines the three LEDs and even the constant current drivers. Modules such as these are commonly strung together in **LED strips**, an extremely popular use of multicolor LEDs and a wonderful way to add colored lighting for status indication, expressiveness, or simply aesthetics to your robot. RGB LED strips typically have four connections: 12V in, and one pin each for R, G, and B. 12V is always connected to a 12V source, and we can turn all the LEDs of one color on by connecting their respective pin to GND, which is exactly what our transistor-based low-side switch can do! The driver chips inside the LED strip take care of controlling the LEDs' current, so we do not need to worry about selecting the appropriate current-limiting resistors. Nonetheless, depending on the length of the strip, the current can become high enough that we need to select capable power transistors for the low-side switches of the three strands. There is no hope that an Arduino microcontroller by itself could handle that much current, or the 12V voltage level. All of this means that driving a multicolor LED strip is a perfect example to put everything that we learned so far about controlling the brightness of LEDs and using transistors to drive them into practice.

For this example, we will be using a bipolar NPN transistor, a *TIP120*, as the low-side switch to switch the current for each color of the LED strip. This transistor can handle 5A continuously, which is sufficient for dozens of very bright LEDs. To turn it on, we just need to connect its emitter pin to GND and provide a small amount of current to its base input, just as in the schematic shown in *Figure 10.2*. The only difference for the multicolor case is that we need three transistors and three Arduino pins to control them. We will use 220 Ohm base resistors to make sure we do not pump too much current into the transistors' bases. The TIP120 has a very high **current gain** of 1,000, which means that 5 mA of base-emitter current provided by the Arduino is enough to switch 5A of collector-emitter load current. To power the LED strip, we also need a 12V source. A three-cell **lithium polymer** (**LiPo**) battery is a good choice for this task or, alternatively, a 12V benchtop power supply. Connect its 12V output to the 12V input of the LED strip and make sure to connect the battery's or power supply's GND terminal to one of the Arduino's GND pins..

The schematic for powering the LED strip is similar to the one shown in *Figure 10.2*, except that this time we need three transistors, and we can omit the resistors to limit the LED current and connect the color wires of the LED strips directly to the collector pins of the transistors. Using this setup, we can control the powerful LEDs along the LED strip just as if they were the tiny LED that is built into our Arduino board. In *Figure 10.5*, the strip is powered from a benchtop power supply and is currently rendering a warm yellow color by mixing light from the green and red LEDs:

Figure 10.5 – A curled-up multicolor LED strip controlled by an Arduino with three power transistors

At this color and brightness, the strip is drawing around 150 mA of current, much more than an Arduino alone could handle but not a problem for our TIP120 low-side switches.

To control the three colors of LEDs with variable brightness via software, we can use the `Dimmer` class that we developed earlier and simply instantiate one instance per color: a `redDimmer`, a `greenDimmer`, and a `blueDimmer` instance on three different pins:

```
Dimmer redDimmer(2, 100);
Dimmer greenDimmer(3, 100);
Dimmer blueDimmer(4, 100);
```

Now, we can control the color of the strip by setting the individual duty cycles of these three dimmers. You can even write a `set_rgb(int r, int g, int b)` wrapper function that takes in brightness levels for all three and uses the dimmers to set them accordingly, or you can write wrappers for certain colors, such as `set_purple()`. To experiment with the color-mixing effect, you can control the colors individually using the input from three potentiometers, or you can use the serial input method that we learned in *Chapter 7, Testing and Debugging with the Arduino IDE*. You will find both of these options in the *LED_strip* program in the GitHub repository. By now, you know plenty of techniques to experiment with this multicolor setup and explore interesting ways to use them. Have fun!

## Other LEDs

Because LEDs are so useful for many applications, there is a large variety of types of LEDs and LED modules. One type that is often useful to create sophisticated lighting effects is the multicolor LED strip with driver chips that are **individually addressable** via a communication bus. This allows you, for example, to create an LED strip, ring, or any pattern, with the ability to control the color and brightness of each LED module individually to implement much more sophisticated light effects than standard LED strips:

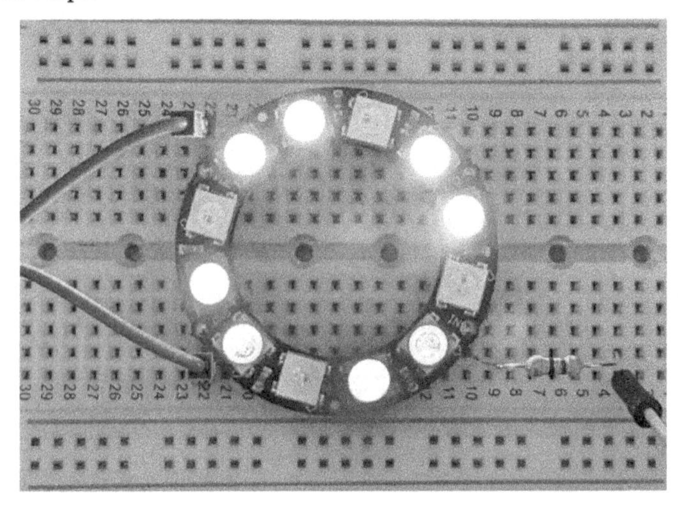

Figure 10.6 – A ring with 12 individually controllable NeoPixel LEDs that can light up at different colors and brightness levels

Adafruit distributes modules like the one shown in *Figure 10.6* under the names *DotStar* and *NeoPixel*, and they also provide Arduino libraries that make them easy to use (see `https://learn.adafruit.com/adafruit-dotstar-leds/overview`).

Another creative use of LEDs is to use them in a two-dimensional array—for example, a 5x5 array of 25 LEDs. Using a clever schematic and one Arduino pin per row and one per column, we can control each of these 25 LEDs with only 10 Arduino pins, one at a time. If we control them in rapid succession, taking advantage of the PoV effect, we can create the illusion of controlling all 25 LEDs simultaneously. This gives us the ability to create a simple display. Taking this one step further, we can stack five of these arrays on top of each other, creating a 5x5x5 LED cube with 125 LEDs that we can control with only 15 pins. This is not really a robotics project anymore, but a fun way to experiment with LEDs, power electronics, graphics, and optical illusions. There are many examples of stunning LED cubes to be found on the internet, including this detailed *Instructable* that shows how to build an 8x8x8 LED cube: `https://www.instructables.com/Led-Cube-8x8x8/`.

# Using different types of displays

To transmit a lot more information between the robot and you (or other people) compared to using LEDs, you can add a display to show text or even graphics.

In this section, we will take a brief look at displays that are suitable for Arduino robots and the trade-offs between several types of them, and we will also step through an example of how to use an LCD with one of the libraries that comes standard with the Arduino IDE.

## Character displays

The most basic display that you can use to display information is the **alphanumeric character display**. The most common type (by a wide margin) is the **Liquid-Crystal Display** (**LCD**). LCDs consist of one or more lines that have several segments each. Each segment can display a standard alphanumeric character (such as an "A" or a "2"), or a user-defined special character such as an emoji. Among LCDs, the 16x2 type with 16 characters per line and 2 lines is the most used type for DIY projects:

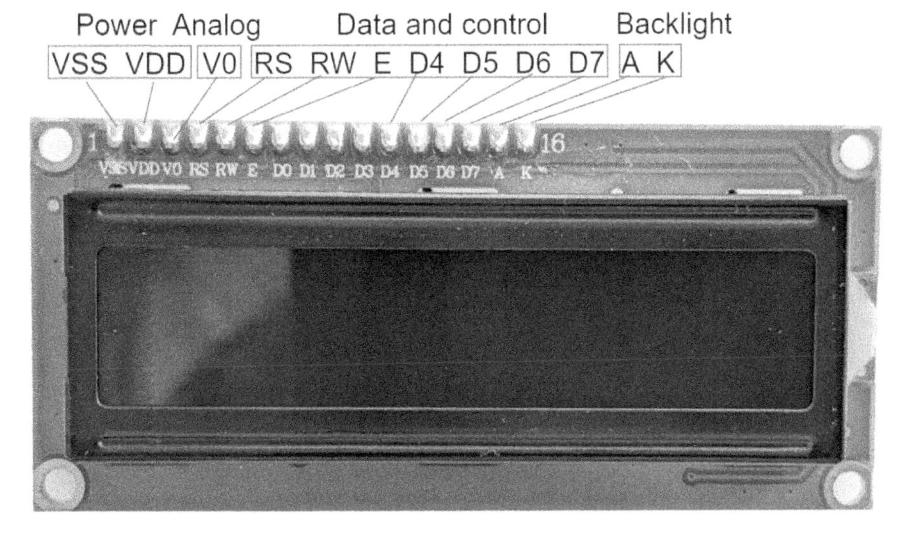


Figure 10.7 – A character LCD on a breadboard-friendly carrier PCB

Controlling the many pixels of an LCD is quite involved, so we need a driver IC that sits between the LCD and our Arduino. This driver does all the heavy lifting of controlling the individual pixels of the display and has a relatively simple interface for our Arduino. Fortunately, LCDs are commonly sold on PCBs that already contain this driver chip, and the interface of these LCD driver chips is standardized. This means the code that we write for controlling any 16x2 LCD will work fine with most other 16x2 LCDs, too. The Arduino IDE comes with the `LiquidCrystal` library built in that uses this standardized interface. Unfortunately, standard LCD controllers do not work with common serial bus interfaces such as **SPI** or **I2C**; instead, instead they use a type of **parallel interface** that requires either four pins (in 4-bit mode) or eight pins (in 8-bit mode) for data transmission. There are quite a few inputs to an LCD controller, so let us break them down into three categories. You can refer to *Figure 10.7* to see where the pins are on the LCD module.

### Data and control pins

For most applications, it makes sense to use 4-bit mode instead of 8-bit mode, thereby saving four pins. Besides the four data pins (*D4*, *D5*, *D6*, *D7*), we also need to drive the LCD controller's Register Select (*RS*) pin, its Register Write (*RW*) pin, as well as its Enable (*E*) pin. This means we need to reserve seven of our Arduino's digital output pins to control a character LCD screen.

### Power pins

To power the driver chip and the display, we need to provide it with 5V on its *VDD* pin and connect the *VSS* pin to GND. LCDs usually have a separate LED for the backlight that we can drive from one of the Arduino's pins and dim with a PWM signal, or we can power it directly from 5V via a 220-Ohm current limiting resistor. The backlight LED pins are either labeled + and - or Anode (*A*) and Cathode (*K*), respectively.

### Analog pin

LCDs even have an analog input pin, usually labeled *V0*. The voltage on this input determines the contrast of the display. To have control over the contrast, we can connect a potentiometer to 5V and GND and connect its output directly to the V0 pin. If you are using an Arduino with a built-in **digital-to-analog converter** (DAC), you can use the analog output to control the contrast programmatically.

*Figure 10.8* shows how to connect a standard character LCD to an Arduino using 4-bit mode:

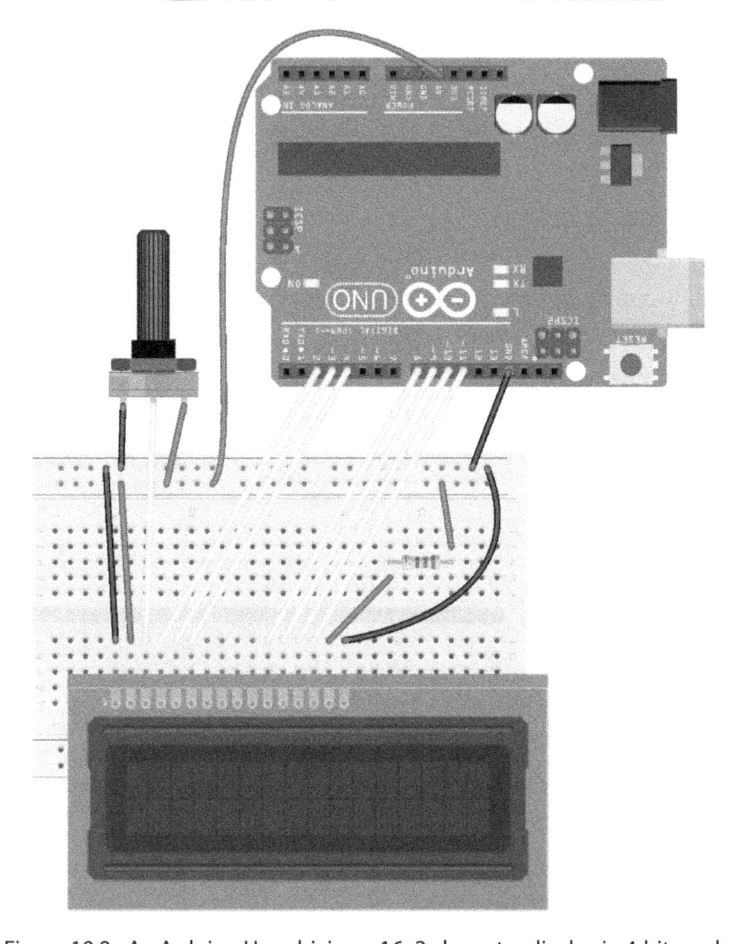

Figure 10.8 –An Arduino Uno driving a 16x2 character display in 4-bit mode

With this setup, the backlight brightness is fixed, but we can control the display's contrast with the potentiometer.

### *Controlling an LCD*

When everything is wired up as described earlier, the Arduino IDE's `LiquidCrystal` library (`https://www.arduino.cc/reference/en/libraries/liquidcrystal/`) makes it easy for us to use the LCD.

> **Note**
>
> Writing on an LCD is cursor-based. This means that anything we write will start at the current cursor location, so we either need to keep track of the cursor location or make sure to set it to the intended position before we write.

As usual, we first include the required library. We then define which pins are connected to which input on the LCD module. We can then create an instance of the `LiquidCrystal` class that we call `myLCD`. We only pass four data pins to the constructor, which tells the library that we are intending to use the 4-bit mode:

```
#include <LiquidCrystal.h>
const short rs_pin = 2;
const short rw_pin = 3;
const short en_pin = 4;
const short d4_pin = 8;
const short d5_pin = 9;
const short d6_pin = 10;
const short d7_pin = 11;
LiquidCrystal myLCD(rs_pin, rw_pin, en_pin, d4_pin, d5_pin,
d6_pin, d7_pin);
```

In the `setup()` function, we call the LCD's `begin()` function to tell it that we are working with a 16x2 LCD. At this point, the cursor will be in the top-left corner, and we can write some text to the first line:

```
void setup()
{
  myLCD.begin(16, 2);
  myLCD.print("Arduino Robots!");
}
```

In the `loop()` function, we set the cursor location to the beginning of the second line (column 0, row 1) and print the current time in milliseconds. We also add a short delay for program stability:

```
void loop() {
  // Set cursor to column 0, row 1.
  myLCD.setCursor(0, 1);
  myLCD.print(millis());
  delay(50);
}
```

In this loop, we continuously overwrite the old output in the second line with a new value. This works well because the new value will never be shorter than the old one. If you instead print the analog value of a potentiometer where the number of digits can decrease, you would continue to see the trailing digits of the old value that did not get overwritten by the new, shorter value. To avoid visible remains of old content, you can either print trailing spaces after new, shorter content or clear the entire display using the library's `clear()` function before printing updated content.

You can adjust the LCD's contrast by turning the potentiometer to get the best display performance under your lighting conditions, and if all went well you should be seeing something very similar to this:

Figure 10.9 – The character LCD in action; the number of wires needed for an LCD approaches the limit of what is practical on a solderless breadboard

Feel free to experiment more with this LCD setup! Some ideas to get you started are as follows:

- Print the analog value of a potentiometer to the display
- Control the brightness of the backlight from your Arduino via PWM

### Controlling LCDs with a serial interface backpack

As you saw, it takes quite a few pins to control an LCD. This can quickly become problematic, especially for more ambitious projects that require lots of pins for other peripherals, or for projects that use a small Arduino board with only a few available pins. To work around this problem, there are backpack boards that add yet another level of interfacing between the actual LCD and the Arduino. The Arduino can talk to these boards using the I2C or SPI interface, and the backpack board creates the required signals for the LCD driver chip. Adafruit sells such a backpack and provides an Arduino library that makes it easy to use. If you ever want to use an LCD but do not have the required number of pins available in your project, using a backpack might be the solution to this problem (`https://www.adafruit.com/product/292`).

## Graphical displays

Character LCDs are a wonderful way to create a user interface for your robot and can make it much more interactive. However, they are still limited in what they can display. There is no way to display images or graphical animations on a character display (ASCII art aside: `https://en.wikipedia.org/wiki/ASCII_art`). To display images and more advanced graphics, we need *graphical displays*. There is a very wide variety of graphical displays available that vary in their underlying technology (mainly **LCD** and **organic LED**, or **OLED**), their resolution (pixel count in either dimension), and their ability to display color or not. There are also graphical touchscreens that can be used with an Arduino. Because of this wide variety, there is less standardization than for character LCDs, and we need to find a matching Arduino library for the specific graphical display that we want to use. *Figure 10.10* shows an example of graphical and text outputs on a tiny 128x32 OLED display with an SPI interface:

Figure 10.10 – A small graphical OLED display with an SPI interface

Graphical displays are more complex and thus more difficult to use than character displays, and they require more memory to store and buffer the content that we want to display. If used well, graphical displays can make your robots look a lot more impressive and professional. If your goal is to build a prototype to bring to trade shows or to present to potential investors or co-founders, investing some time in learning how to use graphical displays effectively might be well worth it.

# Adding sound output to your robot

As with lights and displays, sound can be a great way to let your robot communicate with you or anyone else who might interact with it. Sound is made of sound waves, small pressure waves in the air that propagate to your ears and create the sensation of hearing something. Speakers create these waves by making a membrane vibrate, which in turn makes the air around the membrane vibrate to create sound waves. The pitch of the sound is dependent on the frequency of the vibration. The higher the frequency, the higher the pitch of the sound.

Creating sounds with an Arduino and a simple speaker is very easy. However, there are a few different techniques that you can choose from, so let us look at some of them so that you know your options.

## Digital sound

The easiest way to create sound is by driving a speaker with a square wave, by simply switching the speaker voltage or current on and off at the frequency that corresponds to the pitch that we want to produce. Let us call this type of sound digital sound, as opposed to analog sound, which requires the synthesis of actual analog waveforms.

Digital sound is what most people would describe as buzzing. It is not a very pleasant sound and cannot be used to faithfully reproduce music, but it can be sufficient to let your robot make some expressive beeping sounds (think of R2D2 from *Star Wars*). Arduino even has a standard function called `tone()` built into it (no library needed) to create these square waves with varying frequencies on any digital pin. This function is perfect for driving a buzzer. The simplest buzzer hardware that we can use is **piezo buzzers**. These are small and cheap devices that can be driven directly from the Arduino's pins, without the need for amplifiers or even current limiting resistors. Because they are so small, they are particularly good for producing high-pitched buzzing:

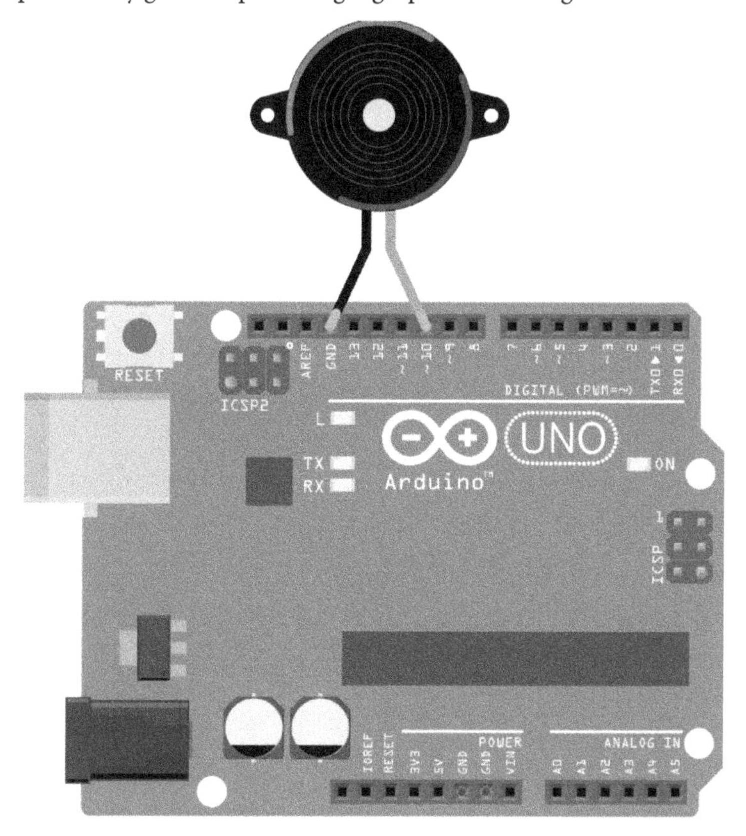


Figure 10.11 – A simple piezo buzzer can be connected directly to the Arduino

You can create sound from a buzzer that is connected to an Arduino output pin and GND with just one line of code: a call to the `tone()` function. This function is non-blocking, so it can continue to produce a tone even while your program has moved on to do other things. Let us look at two short examples of how to have some fun with a piezo buzzer and the tone function.

First, let us map the analog value of a potentiometer to a range of 16 frequencies, and use `tone()` to play the frequency that corresponds to the current value of the potentiometer. Add the following lines to the start of your program:

```
const short analog_input_pin = A0;
const short speaker_pin = 10;
```

And in the `loop()` function, simply add the following:

```
int freq = 100 * ((analogRead(analog_input_pin) >> 6) + 1);
tone(speaker_pin, freq);
delay(5);
```

In the first line, we shift the 10-bit analog value by 6 bits to the right to turn it into a 4-bit value in the range from 0 to 15. We add 1 and multiply the result by 100. This gives us 16 frequencies from 100, 200, 300, ... to 1,600 Hz. In the next line, we use `tone()` to drive the buzzer with this frequency, and we add a small `delay()` function for overall program stability. If you load this program and turn the potentiometer, you will hear a very robotic-sounding range of sounds.

Second, let us write a function that can play a little jingle on the buzzer. At the beginning of the program, we define the frequencies (in Hz) and durations (in milliseconds) of the five sounds in our jingle:

```
int frequencies[] = {440, 523, 659, 523, 440};
int durations[] = {200, 200, 200, 200, 450};
```

Then, we add a function that plays that jingle using `tone()`:

```
void jingle() {
  for(int i = 0; i < 5; i++) {
    tone(speaker_pin, frequencies[i], durations[i]);
    delay(durations[i]);
  }
}
```

This function loops through the entries in the list of frequencies and durations and uses `tone()` to play the specified note, followed by a delay before moving to the next note. In our `setup()` function, we add a line to start the serial interface:

```
Serial.begin(115200);
```

And finally, in our `loop()` function, all we do is check if there is Serial data available. If it is, we play the jingle and then clear the Serial input buffer:

```
if(Serial.available()) {
  jingle();
  while(Serial.available()) {
    Serial.read();
  }
}
```

Now, if you send any data to your Arduino via the Serial Terminal, you will hear the jingle. You may have noticed that this is a simple, blocking implementation of a jingle that makes use of `delay()`. If you want to use this as part of a robot program, you can turn it into a non-blocking state machine version with techniques that we learned earlier. If you are looking for an exercise in implementing non-blocking state machines, this one is for you!

### Analog sound

If you want more than just buzzing, things get a little more difficult. You will need a better speaker that we cannot directly connect to an Arduino pin anymore, so you will need to either build or buy a dedicated audio amplifier. And you will need to drive this amplifier with a smoother waveform than just a simple square wave. Creating these analog waveforms requires a DAC. The Arduino Uno does not have a DAC built in. You can build a DAC for the Arduino Uno either from a network of resistors or by low-pass filtering a high-frequency PWM signal, but then, you also need to create the desired voltage values of the waveform in software at a high frequency. All of this is possible, but it is significantly more complicated than just buzzing with `tone()`, and it rarely adds a lot of value to a DIY robot.

However, if you are trying to create a more polished prototype, it might be worth looking into these techniques. At a high level, you will want to read a sound file from an SD card (stored in the WAV format), decode this file, and use it to drive the DAC that drives the audio amplifier that drives the speaker. You can use the excellent *TMRpcm* library (`https://github.com/TMRh20/TMRpcm/wiki`) as a tool for this purpose, and its documentation as a starting point to learn more.

## Summary

In this chapter, we learned how to drive external LEDs with power transistors and how to control their brightness via software PWM. We saw how to use this technique to render any color and brightness with an RGB LED strip. We also learned about various kinds of displays and worked through an example of displaying text on a character LCD screen. Finally, we learned how to use a piezo buzzer to create beeping sounds and jingles that let your robot communicate its status and be more expressive.

In the next chapter, we will learn how to add a variety of wireless interfaces to your robot, which is especially useful to control mobile robots and collect real-time telemetry data from them.

# Further reading

Besides experimenting more with the examples that we built in this chapter, you can learn more about LCDs, LEDs and how to control them with power transistors from the following resources:

- An in-depth introduction to transistors—for example, for controlling LEDs: `https://learn.sparkfun.com/tutorials/transistors/all`
- An introduction to LEDs: `https://learn.sparkfun.com/tutorials/light-emitting-diodes-leds/all`
- Documentation of the Arduino `LiquidCrystal` library: `https://www.arduino.cc/reference/en/libraries/liquidcrystal/`

# 11

# Adding Wireless Interfaces to Your Robot

If you need to communicate with your robot without a USB cable from a distance, for controlling or viewing live telemetry, your robot needs a wireless interface. There are many wireless interfaces and standards in use. In this chapter, we will look at four that are particularly useful for Arduino robots, from the simple hobby RC interface to full-fledged Wi-Fi.

This chapter is organized into four main parts:

- Controlling your robot with a hobby remote control
- Using XBee to make your serial interface wireless
- Bluetooth control via your phone or tablet
- Connecting to your robot over Wi-Fi and controlling it from any browser

By the end of this chapter, you will know what wireless interface to choose for your project and how to successfully integrate it into your robot.

## Technical requirements

Wireless interfaces require special hardware. Besides your Arduino Uno, a breadboard, jumper wires, and your PC, this is what you need for the examples in this chapter:

- A hobby remote control system consisting of a transmitter and receiver, such as the Radiolink TS8 transmitter (`https://www.radiolink.com/t8s`) and R8EF receiver (`https://www.radiolink.com/r8ef`)
- Two XBee modules, an Arduino XBee shield, and an XBee breakout board with a USB connector

- An Adafruit Bluefruit LE UART Friend module for Bluetooth connectivity (`https://www.adafruit.com/product/2479`)
- An Adafruit HUZZAH32 – ESP32 Feather board for Wi-Fi connectivity (`https://www.adafruit.com/product/3405`)

# What is a wireless interface

We already got to know a lot of different communication interfaces that are commonly used to connect parts of your robot to the Arduino that controls it. These are interfaces such as SPI, I2C, and Serial. All of these have one thing in common, and that is that they transmit data over wires. Therefore, they all fall under the category of **wired interfaces**. In contrast, the interfaces that we will talk about in this chapter transmit data without the need for physical wires – they are wireless.

## When to use wireless interfaces

Not requiring physical wires is an obvious advantage over wired interfaces, as it allows for a lot more flexibility and versatility when it comes to making ad hoc connections with new devices or transmitting data over longer distances. Naturally, you might wonder why we even bother with wired interfaces in the first place. To answer this question, let us look at some of the disadvantages that wireless interfaces have compared to wired ones. In general, these are some disadvantages of wireless interfaces:

- Less reliable
- Need more hardware components
- More expensive to implement
- Need more power
- Have less bandwidth (amount of data that can be transmitted per time)
- Can be subject to local regulations
- More difficult to troubleshoot

As you can see, the unique advantage of being wireless comes with a long list of disadvantages. We therefore only consider using wireless interfaces when the benefit of being wireless outweighs all these disadvantages. This is virtually never the case for components within your robot itself. Any components that are physically on your robot are usually best connected with wired interfaces. However, if your robot is mobile and needs to be untethered, all parts of the system that are not on the robot need to be connected wirelessly. Typical use cases for wireless interfaces are as follows:

- Controlling your robot from a control PC or viewing live telemetry data on a ground station
- Controlling your robot with a wireless **remote control** (**RC**)

- Controlling your robot from your phone or tablet
- Controlling your robot over the internet

We will cover all of these use cases in this chapter. There are others, most notably communication between two or more mobile robots. Some of the techniques we will learn in this chapter can be directly transferred to this application, too.

There is a large variety of standardized wireless interfaces that we can use for Arduino robots. Discussing their underlying technology in more detail is a fascinating topic, but we will instead focus on how to use a practical selection of them – namely, a standard hobby RC, XBee transceivers, Bluetooth, and Wi-Fi in our robotics projects.

## The structure of a wireless interface

All wireless interfaces that we will look at work with radio waves, a certain part of the spectrum of electromagnetic radiation. It is worth mentioning, though, that there are also light-based wireless interfaces (most TV remotes work with infrared light, for example) and even sound-based wireless interfaces (for example, for underwater data transmission). You can think of humans talking to one another as using a sound-based wireless interface for data transmission, albeit a very slow one.

Conceptually, a wireless interface has two parts: a **transmitter** and a **receiver**. The receiver receives the signal that is emitted by the transmitter and decodes the information that was transmitted. Oftentimes, a single device implements a transmitter and a receiver at the same time. These devices are called **transceivers**. A cell phone that allows you to send and receive text messages is an example of a transceiver. Let us start our exploration of practical wireless interfaces with a simple combination of transmitter and receiver, a standard hobby RC system.

# Controlling your robot with a hobby remote control

Hobby RCs have been in use for decades. Even though the underlying radio technology has changed a lot and modern hobby radio controls offer many advanced features compared to the early days of RC technology, the output signal produced by the receiver has remained remarkably simple. This makes them extremely easy to integrate into an Arduino robot.

A hobby RC system consists of a handheld transmitter with control inputs such as sticks, wheels, or buttons, and a receiver. Each manual input to the transmitter is transmitted wirelessly to the receiver and output by the receiver as a digital signal. Between 2 and 10 signals are common among standard systems. The different signals that can be transmitted are called **channels**. With a six-channel RC system, we can control six functions of our robot, for example, six joints of a robot arm.

## Decoding the receiver output

While an RC transmitter can take on many forms, shapes, and sizes, the electrical interface to the receiver is standardized. Thanks to this standardization, there is a lot of hardware, such as standard RC servos and BLDC ESCs that plug right into these three pins and work without any further setup. But we can also use an Arduino microcontroller to connect to the signal pin of any channel to decode the signal that is transmitted on this channel! For each channel, an RC receiver has three pins: GND, 5V, and signal. GND and 5V are connected across all channel outputs, so you can power a receiver over any of its channel ports. The following figure shows an example of an eight-channel RC receiver and its pinout:

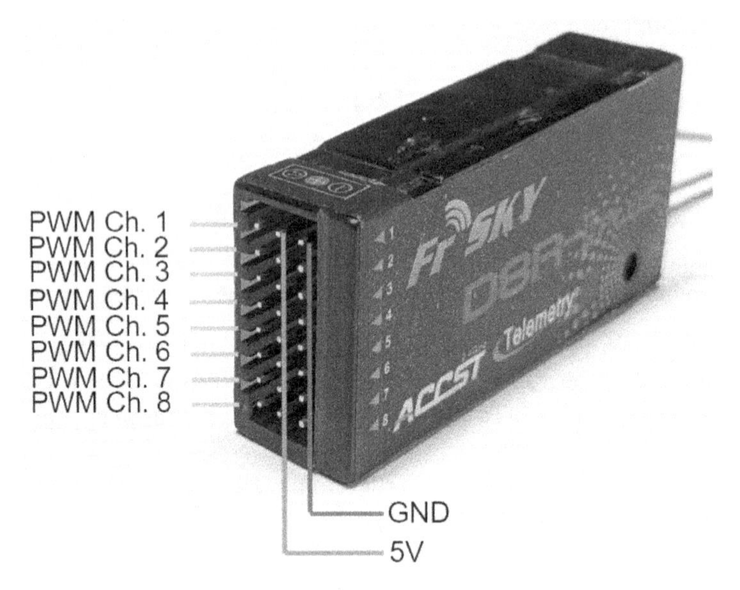


Figure 11.1 – Pinout of an eight-channel RC receiver

We have already learned that RC servos are controlled by a PWM signal, so you might have guessed that the output signals of an RC receiver take the form of PWM signals – more precisely, PWM signals with a *frequency* of 50 Hz and *pulse width* roughly between 1 ms and 2 ms. The information is contained in the pulse width, so to decode the signal, we need to measure the duration of the pulse (see *Figure 11.2*). It is easy to generate this signal with the Arduino servo library, but there is no library built into the Arduino IDE that lets us decode this kind of signal in a non-blocking way. Fortunately, all we need to do is combine our knowledge of the PWM signal with what we have already learned about using *interrupts* to develop the code that lets us decode an RC PWM signal. Let us dive right into a small Arduino program that decodes a single RC channel in a non-blocking, interrupt-driven way.

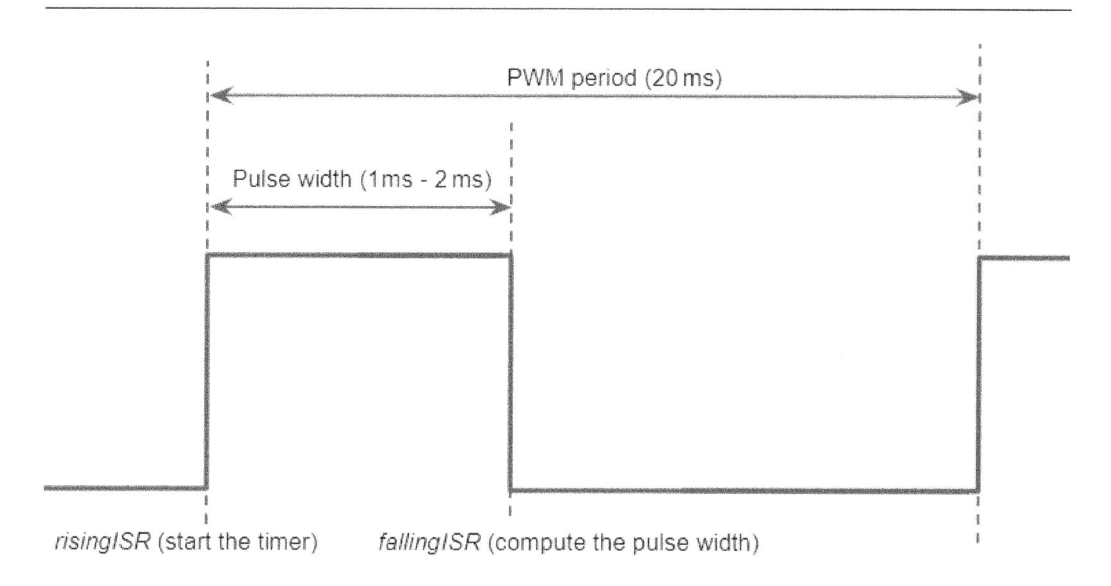


Figure 11.2 – One PWM pulse with the rising and falling signal edges that trigger the interrupts that we use to decode it

At the beginning of the program, we declare global variables to store the last signal rise time and the PWM pulse width in microseconds. We declare these two variables as `volatile` since we will modify them inside interrupt service routines. We also define a `const` variable that contains the number of the PWM input pin:

```
volatile unsigned long rise_time_us;
volatile unsigned int pulse_width_us;
const short pwm_input_pin = 2;
```

In the `setup()` function, we set up an interrupt to catch the rising edge of the signal and map it to an interrupt service routine, `risingISR()`, that we need to provide. We also initialize the serial interface to be able to verify that things are working as expected:

```
void setup() {
  Serial.begin(115200);
  // Attach interrupt handler to rising edge.
  attachInterrupt(
    digitalPinToInterrupt(pwm_input_pin),
    risingISR, RISING);
}
```

In the interrupt service routine for the rising edge, we save the current time in microseconds to the respective global variable. Then, we attach a different ISR – this time, to the falling edge of the PWM input:

```
void risingISR() {
  // Pulse started. Note the start time.
  rise_time_us = micros();
  // Attach interrupt handler to falling edge.
  attachInterrupt(
    digitalPinToInterrupt(pwm_input_pin),
    fallingISR, FALLING);
}
```

In this second ISR, `fallingISR()`, we compute the duration of the pulse as the difference between the rise time and the current time. At this point, we have successfully decoded one pulse of the PWM signal. To get ready for the next pulse, we attach `risingISR()` back to the rising edge of the signal, just as we did in the `setup()` function:

```
void fallingISR() {
  // Pulse is over. Compute pulse width.
  pulse_width_us = micros() - rise_time_us;
  // Attach interrupt handler to rising edge.
  attachInterrupt(
    digitalPinToInterrupt(pwm_input_pin),
    risingISR, RISING);
}
```

With this efficient interrupt mechanism in place, the global `pulse_width_us` variable always contains the duration of the last pulse in microseconds, and we can use this information anywhere in our program. Since this is interrupt-driven, the decoded PWM value will always be accurate, independent of what is happening in the `loop()` function. For our example, we use the `loop()` function only to print the pulse width, and then wait for 20 ms:

```
void loop() {
  Serial.println(pulse_width_us);
  delay(20);
}
```

You can load this code onto your Arduino and connect an RC receiver to it: 5V to 5V, GND to GND, and the signal output of one of the channels to pin 2. When you power up the transmitter and connect the Arduino to your PC, you can see the signal value in the Serial Monitor. This value will range between 1,000 (possibly a little less) and 2,000 (possibly a little more) when you move the input stick or wheel or press the button associated with the channel you are decoding.

The position of sticks, wheels, and sliders usually maps linearly to the pulse width of the output signal. Some transmitters might allow you to change this mapping, but it is usually better to change it in our Arduino code rather than in the transmitter. The state of buttons is usually reflected as either one of the two extreme pulse widths, for example, 1 ms for pressed and 2 ms for released. Sticks are usually self-centering, which is great for inputs that have a logical center, such as the steering angle of a car.

# Replacing the USB cable with a Zigbee module

We have already used the serial interface of our Arduino extensively to communicate with the Arduino IDE's built-in Serial Monitor and Plotter. These tools are incredibly useful to send commands to your robot or visualize live sensor data, controller outputs, or any other data that helps you understand what is happening inside your robot. So far, we always used a USB cable as the communication link between the Arduino and the PC. This is very convenient, particularly because it also works seamlessly for programming the Arduino and providing power from the USB port. But what if we wanted to use the same tools to communicate with a mobile robot that cannot be connected to your PC over a USB cable?

We can easily replace the USB cable with wireless modules that act as the link for the serial interface using a range of technologies. In this section, we will learn how to use Digi's XBee modules for this purpose (`https://www.digi.com/xbee`). They are commonly used in DIY robotics projects, and while they are not the most cost-effective solution for this purpose, they are proven to be reliable. XBee modules are wireless transceivers, sometimes called **radio frequency** (**RF**) transceivers, that can communicate with each other using a protocol built on top of the Zigbee standard (`https://csa-iot.org/all-solutions/zigbee/`). They come in a variety of different form factors, with different antennas, radio power, and functions. But even the most basic XBee modules make it very easy to use them as a transparent wireless bridge for our Arduino's serial interface.

In this example, we will walk through the steps that are needed to use the Serial Monitor on your PC to communicate with your Arduino wirelessly using XBee modules.

## Hardware

We need two XBee transceiver modules: one that is connected to our PC and one that is connected to the Arduino. We also need the hardware to connect them to the PC and the Arduino. We can select from a range of XBees and adapter boards, or we can select a set that contains all these components, such as the XBee wireless kit from SparkFun (`https://www.sparkfun.com/products/15936`). It contains an Arduino shield that fits an XBee module, supplies it with power, and performs the 3.3V to 5V level shifting between the XBee and an Arduino Uno. It also contains a USB-to-XBee board that seats an Xbee and connects to your PC via a standard Mini USB cable.

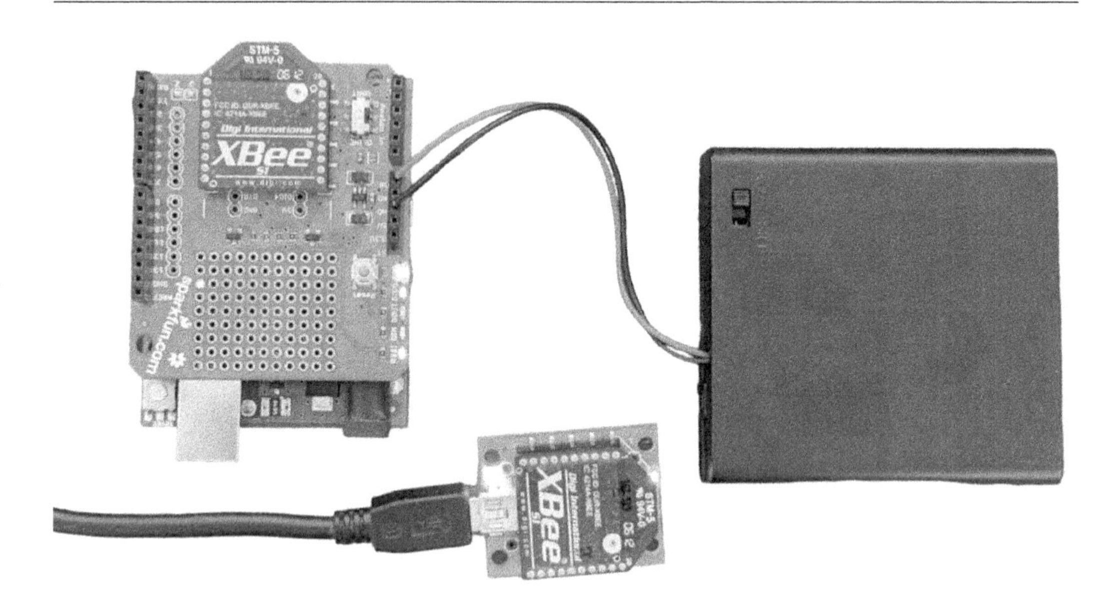


Figure 11.3 – Battery-powered Arduino Uno with XBee shield as well as an XBee on a USB board for wireless communication between your PC and your Arduino

All we need to do is to assemble the XBee shield, insert one of the XBees, and attach the shield to our Arduino. The shield has a little switch on it that determines whether the XBee's serial pins are connected to the Uno's hardware serial (pins 0 and 1, the **UART** switch position) or to pins 2 and 3 for using software serial (the **DLINE** switch position).

## Software

There is no specific Arduino software needed; once you know how to use the XBee modules as a Serial Bridge, you can use them with any program that uses the serial interface. To quickly test and verify that everything is working, we can use a little program that simply lets the Arduino echo any character it receives over the serial interface. In the `setup()` function of this program, all we do is start the serial interface. For this example, we leave the little toggle switch on the XBee shield in the **UART** position:

```
void setup() {
  // Use XBee's default baud rate.
  Serial.begin(9600);
}
```

Normally, we use a baud rate of `115200`. However, the XBee modules' default Serial baud rate is `9600`, so we will use this value this time. To change the XBees' baud rate, you can download and install the free XCTU software (`https://www.digi.com/products/embedded-systems/digi-xbee/digi-xbee-tools/xctu`). You can also use XCTU for troubleshooting and for pairing your XBee modules if you did not purchase them in a set that was already paired. Depending on the specific module and circumstances, XBees work well with baud rates up to `57600`, but not always with our usual `115200`. This is a good example of one of the downsides of wireless interfaces: data rates are generally lower than with wired interfaces.

In the `loop()` function of the test program, we simply check whether a byte was received over Serial, and if yes, we echo that byte back with the `Serial.write()` function:

```
void loop() {
  if (Serial.available()) {
    // Serial echo.
    Serial.write(Serial.read());
  }
}
```

Feel free to first try this program out without the XBee modules attached to verify that it works as expected. Anything you send from the Serial Monitor will be echoed back, character by character. Make sure to set the baud rate of your Serial Monitor to `9600`.

## Using your XBee modules

Once you have confirmed that this works with the USB cable, you can do the same wirelessly. Once the XBee shield and XBee are attached to the Arduino, you need to slide the switch to its **DLINE** position in order to use the Serial port for programming. Once the sketch is uploaded, slide the switch to the UART position. Now, the Arduino's TX and RX pins (pins 0 and 1) are connected directly to the XBee module. You will need to find the serial COM port on your computer that corresponds to the other XBee's USB breakout board. In your Arduino IDE, select that port (choose any board type) and open a Serial Monitor at `9600` baud.

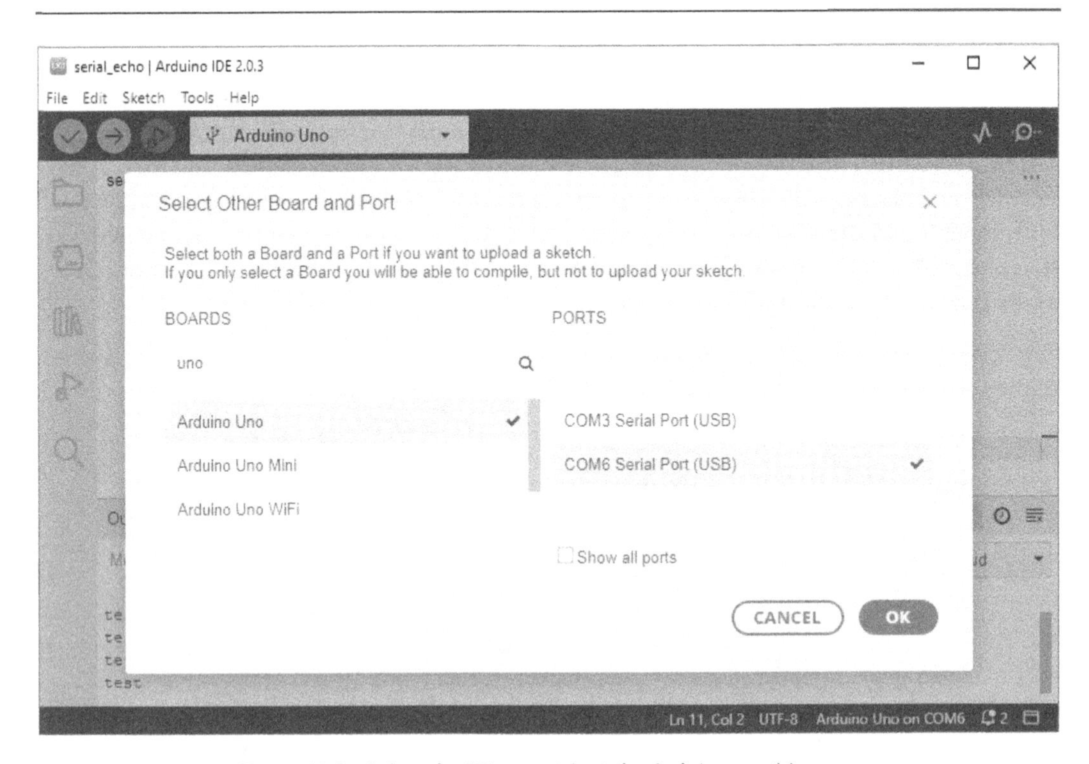


Figure 11.4 – Select the XBee port (not the Arduino port) in your Arduino IDE to communicate with your Arduino wirelessly

Now, everything you send and receive is transmitted between your PC and the Arduino wirelessly! This means you can *disconnect* your Arduino from USB and power it from a battery instead, as shown in *Figure 11.3*, and you will still see the echo from the Arduino in the Serial Monitor. You can experiment with the wireless range by placing your Arduino farther and farther away from your PC and see how far you can communicate with it!

## Other uses

A wireless serial connection to your Arduino robot opens many great and exciting use cases. The Arduino's Serial Monitor and Serial Plotter are great tools for testing and debugging, but you can use a serial connection for many tools outside the Arduino ecosystem. You can even write custom control GUIs and data visualization (called **dashboards**) as well as data logging software for your PC that interacts with your mobile Arduino robot over a wireless serial interface. This can be useful, for example, as a ground station for monitoring the mission of a flying robot or a model rocket. You can also use existing software for this purpose, such as the open source Serial Studio (`https://github.com/Serial-Studio/Serial-Studio`).

# Controlling a robot from your phone via Bluetooth

Another incredibly useful wireless interface that we can use to interact with Arduino robots is **Bluetooth**. Just like Zigbee, Bluetooth is a powerful and very versatile wireless standard that can be used for many applications. We will use it in the most straightforward way, to connect to an Arduino over its Serial interface from a phone or tablet.

## Hardware

This time, one side of the wireless communication will be your smartphone or tablet, so there is no extra hardware required on this end. We only need a Bluetooth adapter for your Arduino. For the next examples, we will use the *Adafruit Bluefruit LE UART Friend* as a **Bluetooth Low Energy** (**BLE**) adapter. This is a versatile and breadboard-friendly adapter with excellent software support. We only need to solder on the pin headers, place it on a solderless breadboard, and connect it to our Arduino, as shown in *Figure 11.5*:

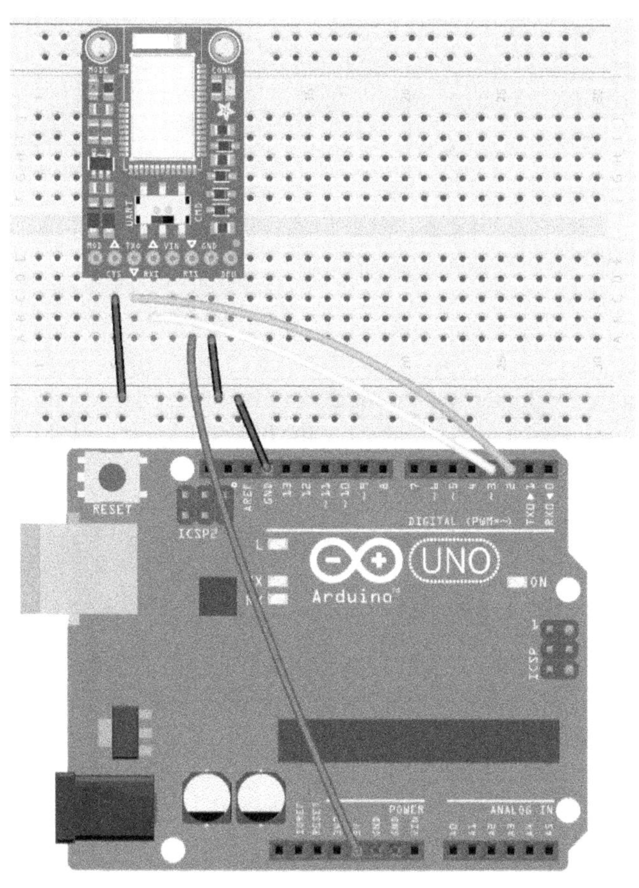


Figure 11.5 – Connecting the Adafruit Bluefruit LE UART Friend to the Arduino Uno

This adapter works on 5V, and its TX and RX pins can be connected directly to your Arduino's digital pins without the need for level-shifting (remember to cross TX and RX). We use pins 2 and 3 for software Serial to leave pins 0 and 1 for hardware serial to the PC. To make this example work, we also need to connect the adapter's CTS pin to GND.

## Bluetooth chat

In our first Bluetooth example application, we will use the BLE adapter to send text messages between the Arduino and your phone. We will use a software Serial port to connect to the Bluetooth module so that we can still use the Serial Monitor via USB. To this end, we first include the `SoftwareSerial` library in our Arduino program and instantiate a software serial port on pins 2 and 3. We then initialize both hardware and software Serial ports in the `setup()` function. Like the XBee modules, the BLE module defaults to a baud rate of `9600`, so we will use this for both ports:

```
#include <SoftwareSerial.h>

// Instantiate software Serial port.
// Pin 2 is for RX, pin 3 for TX.
SoftwareSerial ble(2, 3);

void setup() {
  ble.begin(9600);
  Serial.begin(9600);
}
```

In the `loop()` function, we simply pass through data from one serial port to the other. If data is available from the hardware serial port, we pass it to the BLE serial port, and vice versa:

```
void loop() {
  // Forward Serial data to BLE interface.
  if (Serial.available()) {
    ble.write(Serial.read());
  }
  // Forward BLE data to Serial interface.
  if (ble.available()) {
    Serial.write(ble.read());
  }
}
```

To use your phone or tablet for this experiment, you need to download Adafruit's *Bluefruit Connect* app (`https://learn.adafruit.com/bluefruit-le-connect`). In the app, connect to your Bluefruit adapter and run the firmware update if it pops up. Then, open the **UART** window, which looks similar to the Arduino IDE's Serial Monitor. If your Arduino runs the previous sketch and you have the Serial Monitor at `9600` bps open on your computer, you have created a chat application between your phone and your PC.

Figure 11.6 – The UART screen in the Bluefruit Connect app, chatting with the Arduino IDE's Serial Monitor

Anything you type and send on one end will appear on the other. Congratulations, you just learned how to connect to your Arduino via Bluetooth from your phone or tablet!

## Bluetooth Plotter

The Bluefruit Connect app also has a built-in *plotter* that works just like the Arduino IDE's Serial Plotter. You can use it to visualize live data from your robot right on your phone.

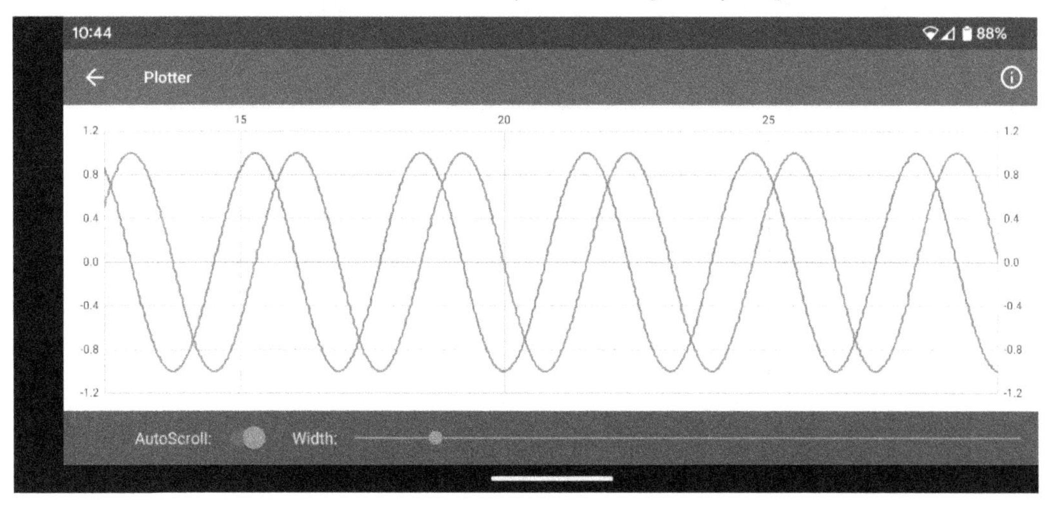


Figure 11.7 – Data visualization from your Arduino right on your phone or tablet

To show a running sine and cosine wave, you simply need to print a line with these two values, separated by a tab, as we would do for displaying it in the Arduino Serial Plotter:

```
void loop() {
  float arg = 2 * millis()/1000.0;
  ble.print(sin(arg));
  ble.print('\t');
  ble.println(cos(arg));
}
```

You can use this plotter to view much more interesting data, such as your robot's battery voltage, motor temperatures, power consumption, or any other value that is interesting to monitor live on your phone while your robot is in operation.

## Bluetooth controller

In this example, we will learn how to use the gamepad controller that is part of the Bluefruit Connect app. If you select **Controller** and **Control Pad** (instead of **UART**) in the app, you will see an 8-button controller with 4 arrows and the numbers 1 to 4. This can be a great tool to control your robot from your phone. If you still run the previous pass-through sketch and watch the Serial Monitor, you will see different sequences of 10 characters appear each time you press one of the buttons on the screen. This is what the app sends to your Arduino when a button is pressed. We can easily find out which button was pressed by focussing on the third character in the sequence, which will be a number between 1 and 8, depending on the button.

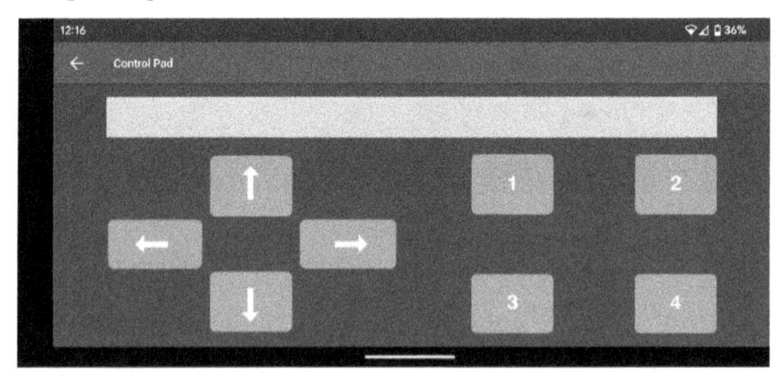


Figure 11.8 – The eight buttons of the Bluetooth Control Pad from the Bluefruit Connect app

Let us write an Arduino program that prints what button was pressed on your phone to the Serial Monitor. The beginning and setup are the same as in the chat example, with the addition of a `const` variable to hold the length of the control pad message and a `char` array to hold the message data itself:

```
const int datalen = 10;
char databuf [datalen];
```

In the `loop()` function, we check whether we received a full command message containing all 10 bytes. If we did, we read the bytes one by one into our message buffer. We then simply check the third byte of the message to decide what button was pressed and print the appropriate message:

```
void loop() {
  if (ble.available() == datalen) {
    // Complete control pad message received.
    // Read the entire message into the buffer.
    for (int i = 0; i < datalen; i++) {
      databuf[i] = ble.read();
    }
    // The third byte tells us which button was pressed.
    if (databuf[2] == '1') {
      Serial.println(1);
    } else if (databuf[2] == '2') {
      Serial.println(2);
    } else if (databuf[2] == '3') {
      Serial.println(3);
    } else if (databuf[2] == '4') {
      Serial.println(4);
    } else if (databuf[2] == '5') {
      Serial.println("UP");
    } else if (databuf[2] == '6') {
      Serial.println("DOWN");
    } else if (databuf[2] == '7') {
      Serial.println("LEFT");
    } else if (databuf[2] == '8') {
      Serial.println("RIGHT");
    }
  }
}
```

If you run this sketch and watch the Serial Monitor output while pressing the buttons on your phone or tablet screen, you will see a printout that corresponds to the button that you pressed. You can easily use this to control your physical robot instead of just printing out what button was pressed. For example, you can add an RC servo motor to this setup and make it move a little to the right each time you hit the right arrow, and to the left when you hit the left arrow, or you can send the servo to predefined poses when you press the number buttons. The possibilities are endless; feel free to try some of them out!

# Connecting to your robot over Wi-Fi

Lastly, we will briefly look at the possibility of connecting to your robot over Wi-Fi. Wi-Fi is an immensely powerful and versatile wireless interface with many potential applications. The way you would most commonly use it on your robot is by running a web server on your Arduino. This server serves up a web interface that you can connect to from any browser over Wi-Fi. It can display live sensor data (even camera data) or expose control interfaces. Wi-Fi is rarely the best choice of wireless interface for your robot due to its complexity. And while there are many great tools available to make using Wi-Fi easy, it is still a lot more complicated to use than any of the other interfaces we discussed so far in this chapter. However, connecting your robot to the internet is great when you need to access it from anywhere in the world – for example, if you build a home security robot that you want to control over the internet when you are away from home. In this case, internet connectivity over Wi-Fi might be the best choice. But keep in mind that any device that is connected to the internet can be hacked, and you certainly do not want a malicious actor to control a robot in your home. This means you will need to familiarize yourself with the basics of internet data security and encryption for a project like this.

A popular domain for Wi-Fi-enabled Arduino projects is the world of the **Internet of Things** (**IoT**). You can find a lot of information and resources on this topic under this search term.

## Hardware

There are Wi-Fi shields for Arduino as well as official Arduino boards with built-in Wi-Fi adapters. But we will take this opportunity as an example of using a third-party Arduino-compatible board, the Adafruit HUZZAH32 – ESP32 Feather board (`https://www.adafruit.com/product/3405`).

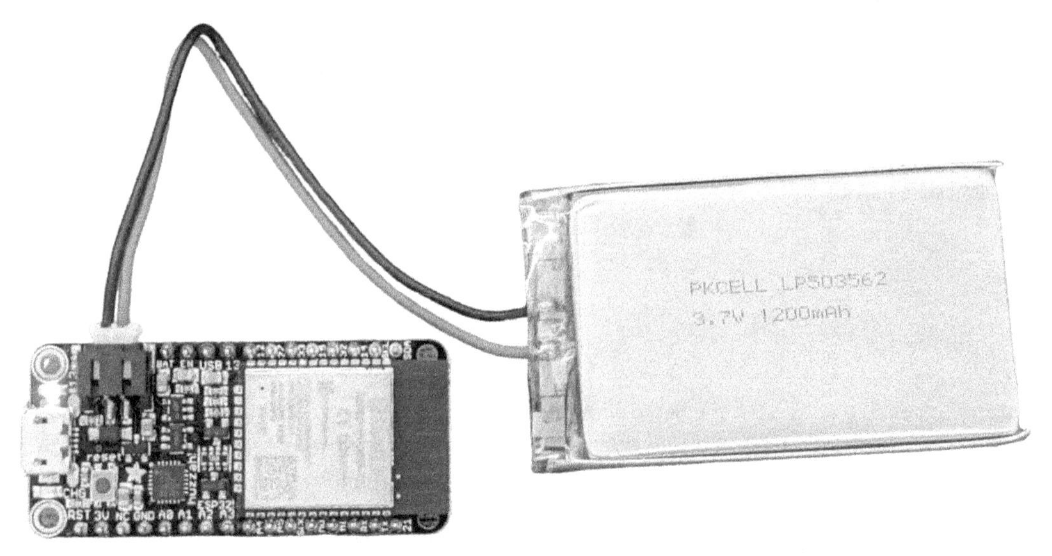


Figure 11.9 – The Adafruit HUZZAH32 – EXP32 Feather board, powered by an external 3.7v LiPo battery

This is, at the time of writing, a cutting-edge development board built around some very powerful hardware components. It has, among other features, a built-in power management system that charges a 3.7V LiPo battery when connected to a USB port, Wi-Fi and Bluetooth interfaces, and a powerful 32-bit microcontroller.

## Setup

To use this board, search for **Adafruit ESP32 Feather** under **BOARDS**. Also, make sure to select the corresponding COM port.

If you cannot find this board in your Arduino IDE, you may need to go to **File | Preferences** and paste the following URL into the **Additional boards manager URLs** box: `https://raw.githubusercontent.com/espressif/arduino-esp32/gh-pages/package_esp32_index.json`.

Then, you can open the **Tools | Board | Boards Manager...** and install the `esp32` package.

To make sure everything works as expected, upload a simple blink sketch:

```
void setup() {
  pinMode(LED_BUILTIN, OUTPUT);
}

void loop() {
  digitalWrite(LED_BUILTIN, HIGH);
  delay(50);
  digitalWrite(LED_BUILTIN, LOW);
  delay(50);
}
```

Compilation of this program, even though it comprises only a few lines, might take longer than what you are used to. This is because your computer needs to compile a lot of background code that allows us to use this powerful hardware just as if it was a simple 8-bit Arduino Uno. If all went well, you should see the red LED blink rapidly after uploading the code. If you have a matching LiPo battery attached to the Feather board, you can unplug your board from the USB port, and it will keep blinking uninterrupted thanks to the onboard power management circuitry.

## Wi-Fi

Let us learn how to control the LED from a web browser instead of having it simply blink. This board has a Wi-Fi interface, and that is what we are here for, after all. The `esp32` board package comes with several example sketches. Go ahead and open the **File | Examples** dropdown. In the **Examples**

**for Adafruit ESP32 Feather** section, select **Wifi | SimpleWifiServer**. This example uses the *Arduino Wi-Fi* library to create a web server that lets you control the board's LED from a browser. But first, we need to make a few modifications to it:

1. Put in your Wi-Fi's SSID as a value for the `ssid` variable.
2. Put your Wi-Fi password as a value for the `password` variable.
3. Replace the number that is used for the LED pin with the `LED_BUILTIN` macro in the call to `pinMode()` in the `setup()` function and in the calls to `digitalWrite()` at the end of the `loop()` function.
4. Optionally, you can change the text that is displayed by the web server to something shorter, for example, `LED ON` and `LED OFF`.

Compile and upload this sketch to your Feather board. Compiling might again take a while; there is a lot of code in the background that needs to be compiled to make this simple web server work.

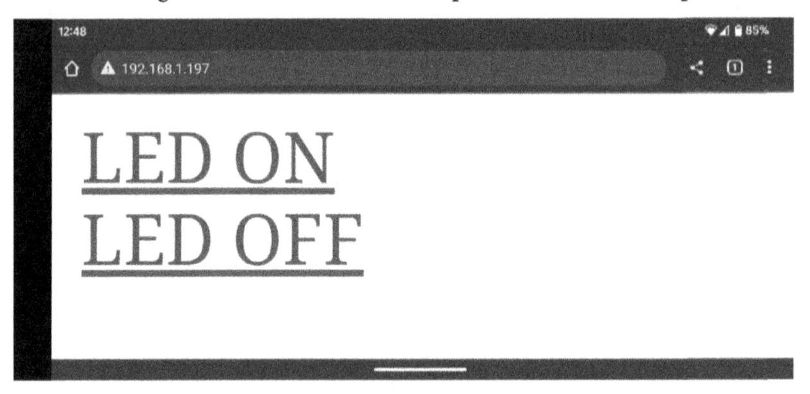


Figure 11.10 – The web page served by the modified SimpleWifiServer example sketch that lets you control the onboard LED

Once the sketch is running, it will connect to your Wi-Fi network and print the ESP32's IP address to the Serial Terminal. You can connect to this IP address (for example, `http://192.168.1.197`) with any web browser on a device (phone, tablet, or PC) that is on the same Wi-Fi network, and you will see a web page that looks similar to the one shown in *Figure 11.10*.

> **Note**
>
> This simple web interface requires an unsecured connection via HTTP. Some browsers might not allow you to connect this way. The ESP32 Feather's hardware also supports encrypted connections, and you can look at the respective example sketches if you want to explore how to use this feature.

Clicking the links should turn the red LED on the board on or off, respectively.

## Next steps with Wi-Fi

We just barely scratched the surface of what you can do with an Arduino board that is connected to your Wi-Fi. Once on Wi-Fi, the Arduino can reach the internet and use the APIs of any website to gather data. For example, your housekeeping robot can query weather data from a weather service to know when to water the plants. Or your robot can subscribe to a certain Twitter feed, and you can send it commands that way. The ESP32 also includes an easy-to-use encryption stack that allows it to easily connect over a secured connection. The possibilities are limitless. A great starting point for further exploration is the other examples that are part of the `esp32` package that you added to your Arduino IDE.

If you think that your robot project will benefit from it, you can also make your Arduino serve a much nicer web page that makes it easier to control your robot from your phone. There are many tools and resources on the web that can help you get started creating a more sophisticated web server with the Arduino Wi-Fi library. There are also several tools that make it very easy to create IoT devices, such as Adafruit's *WipperSnapper* project (`https://learn.adafruit.com/quickstart-adafruit-io-wippersnapper`).

# Summary

In this chapter, we took an in-depth look at several wireless interfaces that you can add to your robotics projects. We learned about their features and what you can use them for. We also went through examples of how to use each of them, and now you have a starting point if you want to add any of them to your robot. Wireless interfaces can be incredibly useful and can make your projects more versatile, but they also add additional complexity and have several downsides compared to wired interfaces.

In the next chapter, we will develop an example robot project that puts many of the lessons we have learned so far to use.

# Further reading

Following are a few helpful resources where you can learn more about XBee modules and the XCTU configuration software, as well as an interesting dashboard project and the Arduino Wi-Fi library:

- A useful XBee buying guide, detailing the difference between many XBee modules: `https://www.sparkfun.com/pages/xbee_guide`
- An introduction to configuring XBee modules with the XCTU software: `https://learn.sparkfun.com/tutorials/exploring-xbees-and-xctu`
- Background information for the Serial Studio dashboard software: `https://serial-studio.github.io/`
- Documentation of the Arduino Wi-Fi library: `https://www.arduino.cc/reference/en/libraries/wifi/`

# Part 4: Advanced Example Projects to Put Your Robotic Skills into Action

This part helps you put your new Arduino and robotics skill into action by guiding you in detail through two exciting robotics projects: a line-following robot that uses a camera to identify a line on the ground, and a self-balancing robot that can carry a smartphone or tablet to provide a fun telepresence experience. These projects put many of the concepts and lessons from the earlier chapters to use, but will certainly teach you even more along the way. This part also contains a chapter with many ideas for the next steps on your journey to becoming an even more proficient roboticist.

This part has the following chapters:

# 12

# Building an Advanced Line-Following Robot Using a Camera

This chapter highlights best practices for building a basic mobile robot base with an Arduino Mega board and geared DC motors. It exemplifies real-time feedback control for autonomous line following with an advanced camera sensor. Building on what we learned in the previous chapters, we will walk step by step through the building and programming process.

This chapter combines a lot of what was covered in the previous chapters and is structured as follows:

- Building a reliable mobile base for this and many other robotics projects
- Wiring and integration of electronic components
- Integrating the Arduino, batteries, and motor drivers
- Writing the code for the line-following application
- Testing and tuning

## Technical requirements

This chapter is a speed walk through the entire process of building a small mobile robot from scratch. You can build your own robot differently with parts you might already have available, but if you want to replicate the exact design, you will need the following components and parts.

- Electronic components:
    - An Arduino Mega2560.
    - A Pixy2 camera sensor. This example uses version 2.1 (for example, from `https://www.robotshop.com/products/charmed-labs-pixy-21-robot-vision-image-sensor-rbc`).

    - An LM298N-based dual DC motor driver board (such as `https://www.amazon.com/dp/B014KMHSW6`).
    - The Adafruit Bluefruit Bluetooth module (`https://www.adafruit.com/product/2479`).
    - A small On/Off switch.
    - A 3-cell, 11.1V LiPo battery that can power your motors (for example, `https://www.amazon.com/dp/B07CQPLC3T`).
    - Two geared DC motors with mounting brackets and wheels (for example, `https://www.amazon.com/dp/B078HYJ3F8`).

> **Important note**
>
> If you are planning to also build the next chapter's self-balancing robot, you might want to get the bigger motors listed in the next chapter instead and build this robot with those.

- Parts:
    - A small solderless breadboard
    - A small caster wheel (for example, `https://www.amazon.com/dp/B09P4729NZ`)
- Tools and supplies:
    - 2 ft of 2 X 3/4" timber (construction grade)
    - Cable ties
    - Jumper wires
    - Basic tools: wood saw, sandpaper, and screwdrivers
    - Wood screws of different sizes
    - A 3D printer (not totally necessary)

You will also need a smartphone with built-in Bluetooth to control the robot.

## Line following – the "Hello, World!" of robotics

Building a mobile robot that can follow a dark line on a bright background is a great project to put much of our learning to use, without being overly complicated. By the end of this chapter, you will have a robot that can autonomously follow a line, completely on its own. You will see feedback control, DC motor control, and Bluetooth communication in action, all in one project. Even though this robot is not complicated, and we will only demonstrate a very basic line-following algorithm, it gives you a great platform to experiment with many more ideas. You can improve the line-following algorithm to make it go as fast as possible or you can add additional sensors to give your robot more capabilities. You can add a camera gimbal so that your robot can use the camera for more than just line following. You can develop a ground station to view live telemetry from your robot, or you can add an RC receiver and add an RC car mode. Once you have a basic mobile robot, there is so much you can try out and learn!

No matter what programming languages you have learned in the past, the chances are that you have written a *"Hello, World!"* program. This is a program that simply prints out its namesake message, and it is routinely used to verify that your programming environment is set up correctly and that you have a good enough grasp of the language and technology to write and run a basic program. In robotics, the bar is a little bit higher since programming is only one part of it. You also need to mechanically build a robot and integrate the necessary electronic components. Building a mobile robot that can autonomously follow a line is a wonderful way to test all these skills and is, therefore, an important milestone in your journey to becoming an expert roboticist. It is a popular project for high school and college students and exposes you to many of the technical challenges of robotics, without being overwhelmingly complex. The following photo gives you an idea of what exactly we are going to build.

Figure 12.1 – The mobile line-following robot that we are going to build

The robot that we will develop in this chapter is a reliable, capable, and extensible example of an Arduino-based line follower. What makes it much more capable than most other line followers you might see elsewhere is its choice of line sensor: it does not use the common setup of one or more discrete phototransistors with their extremely limited resolution. Instead, it uses a camera that provides the robot with much better line detection performance. Not only can the camera detect a wider variety of (very thin) lines under a wide range of conditions but it also provides much more data that you can use to develop better line-following algorithms. It can detect intersections and even read barcodes along the way. The downside of the camera is its price tag. It is considerably more expensive than a few photo resistors, but it can be worth the extra expense, given the opportunities for further experiments and learning that it provides.

# Building a reliable mobile base

A simple and reliable mobile base for a robot is an incredibly useful thing to have as you start building your own, increasingly complex robots. There are many good designs and starter kits available for this purpose, but to solidify what we have learned about building robots, we will build ours from scratch. The design is focused on simplicity, low cost, and flexibility. It has two driven wheels and a passive caster, and it will serve you well as a base for many further experiments and robot designs.

## Building the frame

The T-shaped frame is made of nothing but two pieces of construction-grade wood, cut to length. We will refer to the part that forms the top of the T as the front, but for other applications, that might just as well be the back. Construction-grade wood is cheap and sturdy, as well as easy to source. It is also easy to work with using common tools, and it makes it easy to attach boards, batteries, sensors, and other additions with screws, tape, or glue.

The front part holds the motors underneath it and the Arduino board on the top. It also serves as the attachment point for the camera that senses the line. The length of this bar is driven by the length of the motors that you want to use to drive the wheels. It should be a little longer than the two motors combined, to make sure they fit underneath without problems.

The part of the frame that forms the stem of the T is made from the same material. On the top, it seats the motor controller board and a solderless breadboard, as well as the robot's main power switch. The robot's battery is attached to the underside of this part, and the caster wheel is attached to the rear end of it. The length of this part is driven by the size of the motor controller board and the length of the breadboard to make sure both fit on top easily.

The two best ways to join the two pieces and form the T shape are either with wood glue or with screws. Using glue is the easier option since you do not need any additional tools, other than a vice or C-clamp for clamping the parts together while the glue is curing. Using screws, on the other hand, allows you to disassemble the robot and swap out the wooden parts for versions with different lengths if you need to.

After you cut the pieces to length with a saw, lightly sand them with 120 grit sandpaper to break sharp edges and remove any splinters. Place the two parts together such that they form a T. The two pieces should overlap a little, leaving sufficient space for the motors under the front bar. With the pieces held in place, mark the position of the rear part on the front part with a pencil before you go ahead and attach it more permanently with screws or glue.

If you use screws to join the two pieces, make sure to pre-drill the holes with appropriately sized drill bits to prevent the wood from cracking. Three screws are all that you need.

If you use wood glue, apply and spread the glue to both pieces and then clamp them together with a C-clamp, a vice, or a heavy weight for a few hours for the glue to set.

## Adding motors and wheels

So far, all we have are two pieces of wood joined together – not exactly a robot yet. Let us add motors and wheels as the next step toward turning the frame into a mobile robot. This robot will have two different types of wheels: two driven front wheels and a passive caster wheel at the back (see *Figure 12.2*).

Figure 12.2 – Underside of the wooden robot chassis with motor and wheels at the front, and a passive caster wheel at the back

There are several motor types to choose from (DC, BLDC, stepper, or servo). We will use geared DC motors for our robot, primarily because they are so easy to control. In addition, they are easy to integrate mechanically, very robust, and a cost-effective way to power a small mobile robot. You can find a variety of geared DC motors in robotics online shops such as Pololu, Adafruit, Sparkfun, and even Amazon, making them easy to obtain. You will want to be looking for a 12V DC motor that is between 25 mm and 40 mm in diameter, with a gear ratio in the neighborhood of 20:1. Even though we are not going to use motor encoders for our simple line follower, it is an excellent choice to opt for motors with encoders so that you have them available for future projects when you might need them. We also need the hardware to mount the motors to the wooden frame. Geared DC motors often come with mounting brackets, or you can buy the brackets as additional accessories. Alternatively, you can design your own mounting brackets in CAD software and print them on a 3D printer. Lastly, we are also going to need wheels that connect to the output shafts of the gearboxes. Ideally, you buy the motors, mounting brackets, and wheels all as one kit. Alternatively, you can buy the wheels separately, making sure that the diameter of the wheel hub and the motor shaft match.

Once you have all the parts for your robot's drive train, place the brackets in the front corners on the underside of the front part of your robot frame and mark the locations of the mounting holes with a pencil. Drill pilot holes in these locations and attach the brackets to the frame with wood screws. Then, attach the motors to the brackets with fasteners that fit the motors. Attach the wheels, and the drive system is complete.

The caster wheel at the rear is there to support the back and to allow the robot to steer in any direction without impediment. You should try to find one that has the right size such that your robot is level with all three wheels attached. You can find caster wheels on Amazon or in your local hardware store. Once you have found the right wheel, place it centered under the rear end of the rear part of the robot frame and mark the locations with a pencil. Drill pilot holes and attach the caster wheel with wood screws. With the wheels attached, we are a lot closer to a real robot already!

## Electronic components

With the mechanical build of our simple mobile robot platform completed, we need to integrate the required electronic components. Here is a list for reference:

- Arduino board (preferably an Arduino Mega2560)
- LM298 motor controller board
- On/off switch
- Small, solderless breadboard
- Bluetooth adapter (Adafruit Bluefruit)
- Battery

You can refer to the parts list under the *Technical requirements* section at the very beginning of the chapter for specific products.

### *Integrating the Arduino*

First, we need to choose our Arduino board. An effective way of going about this in practice is to start with the question *Can I use an Arduino Uno?*. If there is no reason not to use an Uno, the Arduino Uno is an excellent choice due to its robustness and breadth of available accessories and support. For our simple line follower, an Arduino Uno is indeed all we need to start with. The camera (which we will look at in more detail a little later) only needs an SPI interface, and we can use Software Serial to interface the Bluetooth adapter. However, there are two arguments against using the Uno, which come from the desire to make this robot platform extensible. For one, Software Serial uses substantial memory and CPU resources of the Arduino Uno. If you want to try out more advanced projects on this robot, you may run into unexpected performance issues caused by Software Serial. Thus, *a board with an additional Hardware Serial port* is preferable. In addition, the Uno cannot handle both motor encoders at full resolution due to the lack of interrupt pins, since reading a quadrature encoder signal requires two interrupt inputs per motor. It is possible to read a quadrature signal with a single interrupt input, but that way, we can only achieve half the resolution and put a lot of burden on the CPU when the motor is spinning fast. For that reason, a *board with at least four interrupt pins* is better suited for our robot platform. Since we are not constrained in size or power consumption, the best choice for the Arduino board is the Arduino Mega, so we will use this as the baseline board in this chapter.

However, since the Arduino Uno fulfills all the necessary requirements for this project, you can just as well use an Arduino Uno if you do not have an Arduino Mega on hand yet. We will make sure to write the code in such a way that it works for both boards. We will use this as an opportunity to learn more about using the **C++ pre-processor**. So far, we have only used it without further mentioning it to include header files with the `#include` directive, but it can do many more useful things for us.

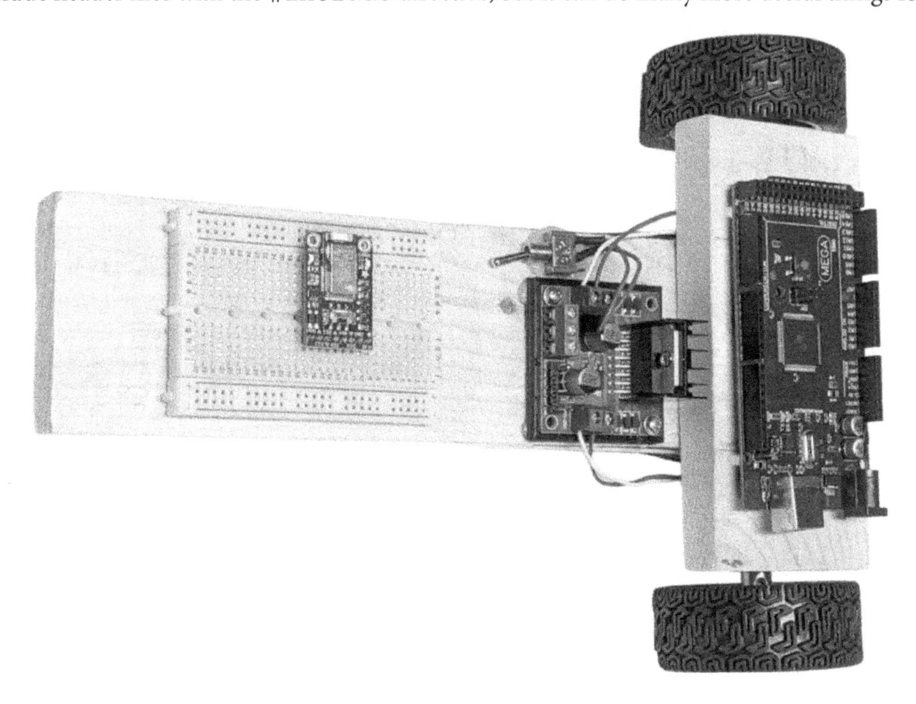


Figure 12.3 – The robot chassis with all electronic components, except the battery

The Arduino Mega fits nicely on top of the front part of the robot's frame. If we place it with the USB connector facing to the right side, most of the pins that we will be using are in the back, keeping the required jumper cables to the motor driver and the breadboard as short as possible, and the overall cabling well organized. Options for attaching the board to the robot include screws, hot glue, and foam-core double-sided tape. To use screws, you will need spacers between the board and the wood, otherwise, you risk bending or even fracturing the board when you tighten the screws (genuine Arduino boards come with acrylic spacers). Hot glue is best regarded as a last resort option, due to its permanence and tendency to get a little messy. That leaves us with double-sided tape as the simplest option for mounting the Arduino to the wood. Depending on the thickness of your tape, you might need more than one layer to make the tape thicker than the little pins that stick out the back of the Arduino board. Using tape also has the advantage of acting as a predetermined breaking point – for example, when your robot starts driving but the USB cable is still attached to your PC. Instead of damaging the board or your PC, the tape can simply sheer off, leaving all the components undamaged.

### The motor driver

We need a motor driver that can handle two DC motors (a so-called dual channel motor driver). The standard solution for this application is to use the ST LM298N chip that we have encountered before, both on the official Arduino motor shield and on the more generic board that we used for the DC servo motor project. For this chapter's robot, we will again use a driver that is built around this chip. Because it is such a common component, there are many vendors that sell LM298N-based motor driver boards online, including Amazon. What is nice about many of these boards is that they often come with an integrated 5V regulator, such as the ST L78M05, which provides up to 500 mA of current. We can use this built-in regulator not only to power the motor driver itself, but all the electronic components of our robot that need 5V, including the Arduino board. These motor driver bards usually either have a switch or a little jumper that determines whether the onboard 5V regulator is connected to the battery or not. If we use the regulator output to power our robot's logic components, this switch becomes the primary power switch of our robot, which is very handy. It is important to keep in mind, though, that this switch only disables the 5V rail of the robot; *it does not disconnect the motor driver from the battery voltage*. A very small amount of current might still flow from the battery, even if the robot is powered off, so it is good practice to disconnect the battery from the motor driver when you are done using your robot.

The motor driver goes on the top of your robot, just behind the front bar. To simplify wiring, place it with the heat sink facing forward. To mount it to the frame, the same considerations as for the Arduino board apply. The only difference is that hot glue is not an option for this board since it might heat up beyond the melting point of hot glue. That is because both the LM293N driver chip and the L76M05 voltage regulator handle large currents and can heat up, accordingly (hence the heat sink that you commonly find on these boards). There is no need for a predetermined breaking point for this board, so using screws is the best option here. You will need spacers under the mounting holes that are higher than the pins that stick out on the bottom side. An effective way to prop up the board is to design and 3D-print a little frame that the board can sit in, and that holds it high enough above the wood surface to let the pins clear the wood.

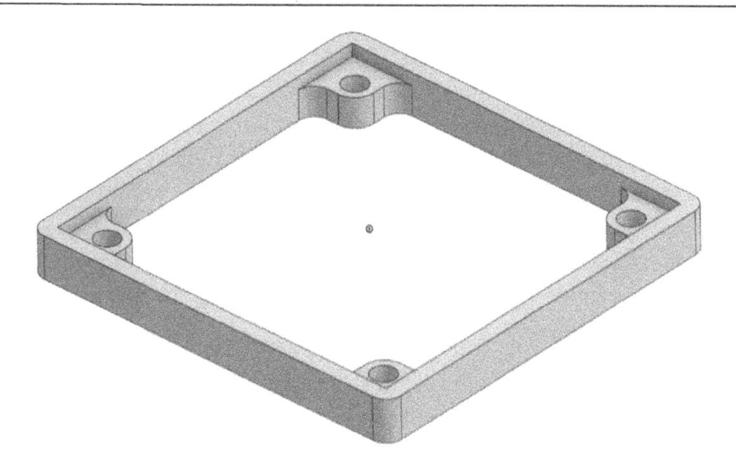

Figure 12.4 – A straightforward, 3D-printer-friendly design for a mounting frame that holds the motor controller

Once you have the frame or other spacers ready, put the board in its intended location and mark the location of the mounting holes with a pencil. Drill pilot holes and secure the board with four mounting screws.

### The power switch

If your motor driver board does not have a built-in switch but instead two pins connected with a jumper, you can use an external switch. Since the current on the 5V level will be small, almost any type of switch will be sufficient. Solder two jumper wires with female ends to the switch and insulate the solder joints with a shrink tube. Now, you can replace the 5V jumper on the board with the switch. This lets you control the 5V output of the motor driver board, and the switch is now the on/off switch for your robot.

The best way to mount the switch depends on the type of switch that you are using and its housing. In this example, we use a rocker switch that is designed for panel mounting. It does not have mounting holes, and the required actuation force is so high that double-sided tape would not be a strong enough way to mount it. Hence, this is a great application for hot glue. When you mount anything with hot glue, especially parts that contain metal like this switch, you need to ensure that the hot glue gun is fully heated up. If the glue is not hot enough, it will cool rapidly in contact with metal and does not form a good bond. Always use caution when using hot glue to not get molten hot glue in contact with your skin!

### Breadboard and Bluetooth adapter

Adding a solderless breadboard to your robot base has several advantages. The biggest advantage is that it allows you to add additional hardware very easily, from sensing resistors to LEDs and breadboard-friendly breakout boards – such as the Bluetooth adapter board that we will use for this project. Having an onboard breadboard is a big part of what makes this robot such a great platform for many future projects of yours. The breadboard also serves as a convenient hub for power distribution, with its two sets of power rails along the two long edges. We will use both rails marked with a - sign to GND, one of the rails marked with a + sign for 5V, and the other one for the battery voltage. This makes it easy to connect other components that you might want to add later to power and GND. Solderless breadboards often come with double-sided adhesive tape already installed on their back side, so all you need to do to attach it to the robot is peel off the protective film and stick it on. On this robot, the breadboard is placed toward the rear end of the top side of the rear wooden part of the frame.

We use the same Adafruit Bluefruit Bluetooth adapter that we got to know about in the previous chapter. It sits directly on the breadboard, and you can use short jumper wires to connect it to GND and 5V.

In addition, we add two resistors connected in series between the battery voltage rail and GND. These resistors form a voltage divider that enables the Arduino to measure and monitor the battery voltage. To choose the values for these resistors, let us recall how a voltage divider with two series resistors, R1 and R2, between a voltage U0 looks and works.

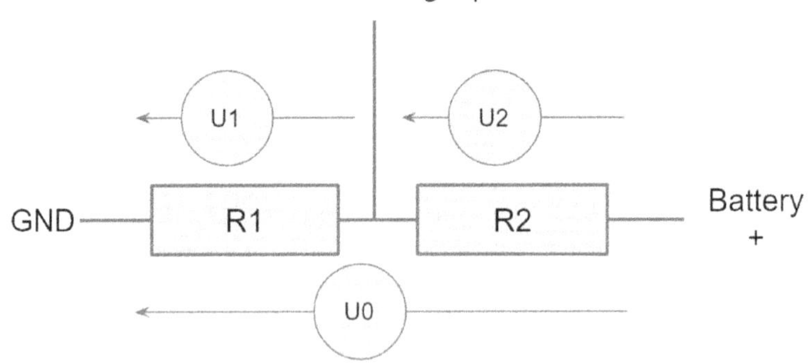


Figure 12.5 – A voltage divider that lets the Arduino monitor a battery voltage that is higher than 5V

We assume that R1 is the low-side resistor (connected to GND) and R2 is the high-side resistor (connected to the battery voltage). The voltage U1 across R1 is the ratio of R1 to the sum or R1 and R2, multiplied by the voltage U0:

$$U_1 = U_0 \frac{R_1}{R_1 + R_2}$$

In effect, the voltage divider scales the battery voltage U0 down to a voltage U1 that we can safely feed to the Arduino Uno's analog input pins. And as long as we know the resistors' values we can easily reverse this process in software to compute the original battery voltage.To scale the voltage of a 12V battery down to 5V or less, R1 should be around half of R2. This results in a scaling factor of around 0.33. Other than that, we simply want these resistors to have resistance values in the kOhm range to avoid unnecessary battery drain. There are many values you can choose that meet these requirements.

For this example, we choose R1 as 4.7 kOhm and R2 as 10 kOhm. The resistors plug right into the breadboard, making their integration quite easy.

### *The battery*

To complete the set of necessary components for the robot base, all that is left to add is the battery. This means we need to decide which battery type to use as well as the voltage and capacity.

We want the battery voltage to be somewhere between 7V and 12V for the 5V regulator on the motor driver board to function optimally without getting too hot. And we want at least 1,000 mAh of capacity to get good runtime without needing to recharge the battery during experiments.

We should not use non-rechargeable alkaline AA batteries or 9V block batteries. Even if they have the appropriate voltage, their current capacity is not sufficient to power two DC motors without dropping their voltage significantly under load. Alternatively, we can use six or eight rechargeable NiMH AA batteries wired in series, using a battery holder. This will yield the appropriate voltage between 7 and 12V as well as the sufficient current capacity to power the motors. The other viable alternative is to use two or three rechargeable lithium battery cells in series. Lithium batteries have a higher energy and power density than the other battery types. While this is a useful feature to enable the use of powerful motors and long battery runtimes, it comes with the added dangers of high-power batteries that we already discussed. Both choices, NiMH and lithium, are good ones for this project. In the example, we will use a three-cell **lithium polymer** (**LiPo**) battery. These batteries are cost-effective, have an excellent energy density, and make our robot design future-proof in case you ever want to upgrade to more motors or more powerful motors. However, we need to be aware of the destructive potential of the extremely high currents that these batteries can provide and thus be careful when working with them. You will also need a charger that is suitable for LiPo batteries.

The battery is placed under the rear part of the robot frame. There are a number of ways you can attach it to the robot, depending on the size and weight of the battery that you choose. If it is a lightweight battery, self-adhesive hook-and-loop tape can be the ideal solution for easy attachment and detachment. For a heavier battery, the hook-and-loop tape might not be strong enough. Another simple and effective option is to secure the battery with two cable ties. The price for this simplicity is that you will need to cut and replace the cable ties every time you want to take the battery off the robot. Alternatively, you can use reusable cable ties that have latches that allow you to unlock them, or hook-and-loop cable wraps to make battery swaps easier.

## Wiring up the robot

With the robot built and the electronic components attached, it is time to wire it up! The importance of good wiring of a robot is an often-overlooked part of a robot build, but it is critical to your robot's reliability. *Figure 12.6* gives you an overview of the complete wiring that we will discuss in detail in the following paragraphs:

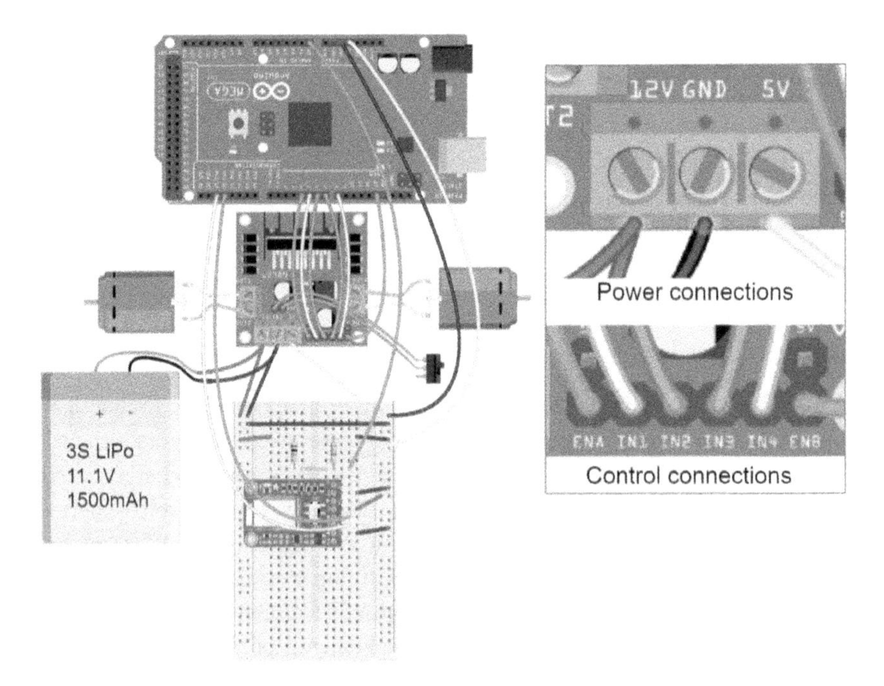


Figure 12.6 – Symbolic wiring diagram of the line-following robot with close-ups of the motor driver connections

Let us split the aspects of wiring into three categories (battery and motor, connections on the breadboard, and connections between boards) and go over these three categories one by one.

## Battery and motor connections

Most hobby lithium batteries come with high-current connectors, such as the red *Deans* connectors (also called T-connectors), or the yellow *XT60* connectors. These connectors are designed to carry high currents and ensure correct polarity when you mate them. You will need to buy or make a mating pigtail connector cable that you can use to connect the battery to the motor controller board. It is a very good idea to integrate a 5A fuse into this cable, especially if you use lithium batteries. You can connect the pigtail to the motor controller via the screw terminals on the board, which makes for a secure and safe connection. Just make sure that there are no stray copper strands coming from your cable that could potentially cause a short circuit between the battery poles (you can tin the cable ends with a soldering iron or use crimp-on ferrules to eliminate this risk). Also, double-check that the polarity is correct. Because the cables from the batteries carry the most current anywhere in your robot, you will want to use thick cable stock for this connection. Soldering thick cables to high-current connectors requires a strong soldering iron that can deliver the required heat quickly. If your soldering iron is not strong enough and takes a long time to heat up the cable and connectors, you might damage both and end up with an unreliable solder joint.

Most L298N motor driver boards have screw terminals for the motor connections as well. Attach cables to your motors with the included connectors or by soldering directly to the motor terminals. If your motor has an encoder and the motor and encoder cables all exit in one bundle, consult the datasheet to identify the two motor cables. Connect them to the two outputs of the motor driver shield via the screw terminals. Since we do not use the motor encoders for this project, you can just stow them somewhere – for example, by routing them along the frame of your robot and securing them with clear tape.

## On-breadboard connections

So far, we always used flexible jumper cables to connect our Arduino board to other parts of our project. There is a better solution for connections between components and rails on a breadboard, namely, rigid jumper wires. You can buy them in sets of different lengths and colors. They are the best choice for on-breadboard connections because they keep the board much more organized than long, flexible cables. For this robot, you can use these wires to connect the two ground rails along the outer edges of the breadboard and to connect the Bluefruit BLE adapter to power. You can also use this type of wire to make the connections between the GND, 5V, and battery voltage terminals of the motor controller and the respective power rails of the breadboard, just to keep the wiring as organized as possible.

## Inter-board connections

For the connections between the three boards, we can use the same flexible jumper cables that we have been using in the previous experiments. None of these connections carry significant currents, so thin jumpers are sufficient. Battery power, GND, and 5V need to be routed from the motor controller to the breadboard where we use one of the power rails for the battery voltage and the other for 5V. From there, 5V and GND are connected to the Arduino board. We also connect the Arduino's Serial2 port to the Bluefruit BLE adapter on the breadboard (remember to cross TX and RX) and the Arduino's A0 analog input to the center point of the battery voltage divider. Lastly, three signals per motor need to be connected from the Arduino to the motor controller: one for enable input (PWM) and two for the direction inputs.

> **Note**
>
> It is important to ensure that the GND inputs of all components of the robot are connected to one another. Otherwise, potentially confusing and unexpected behavior can occur.

The following is a photo of how the robot looks when it is fully wired up:

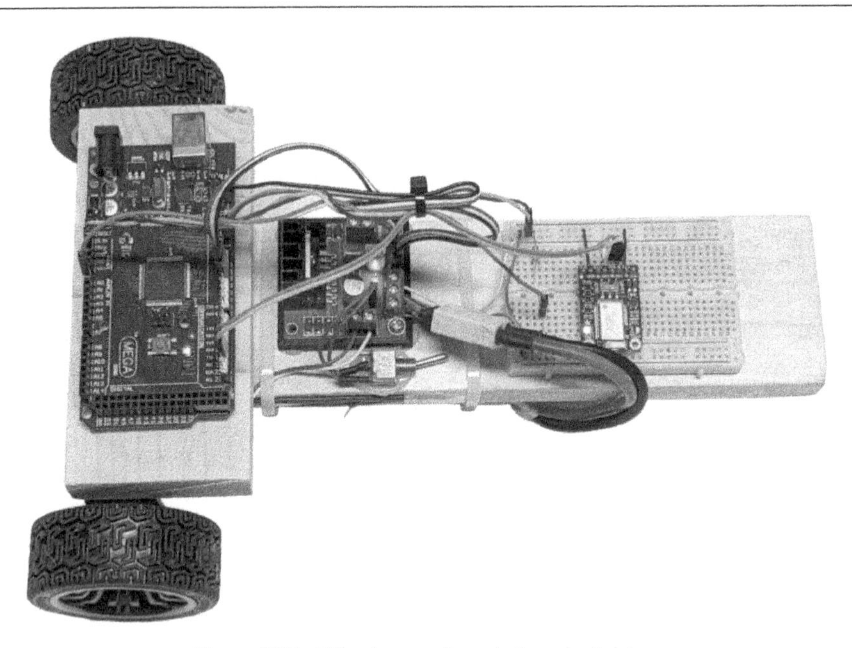

Figure 12.7 – Wired up and ready for a test drive

If you build the robot with the bigger motors and wheels that we will be using for the next chapter's self-balancing robot, it will look like this:

Figure 12.8 – The same robot, but with bigger motors and wheels (and with the camera attached)

With the wiring done, you have completed your first mobile robot platform, which you can now use as the base for a lot of exciting experiments. Congratulations!

# Integrating the camera

Now, let us turn your mobile base into a robot with a camera that can follow a line. To that end, we need to integrate one more component – the camera. Cameras are not a very common sensor for DIY Arduino robots since they produce a lot of data that is difficult to process on a small microcontroller. To circumvent this problem, the PixyCam modules (`https://pixycam.com/`) do all the image processing on board and let the Arduino access very high-level information, such as the location of color patches or the beginning and end of a line in an image. A PixyCam makes for a very powerful line sensor, and you can use it in many other interesting projects, such as sorting machines, object trackers, or even an automatic Rubik's Cube solver. We will be using *Pixy2.1*, but other versions will work just as well.

## Electrical integration

The Pixy camera runs off 5V and can communicate with the Arduino board over SPI, which makes the electrical integration easy. Even better, it comes with an adapter cable that plugs right into the Arduino's ICSP port which carries GND, 5V, and an SPI interface. All we have to do is to plug one end of this cable into the Pixy, and the other into the Arduino's ICSP port.

> **Note**
>
> The PixyCam cable must be plugged into the ICSP port the right way around to avoid damage. Plug it in such that the flat cable exits toward the outer edge of the Arduino board, not toward the board's center.

## Mechanical integration

The camera needs to be in front of the robot, just a few centimeters above the floor, looking down. Before you mount the camera, you can use the PixyMon software (`https://pixycam.com/downloads-pixy2/`) to see a live view of the camera image on your PC and determine a position that gets a good, narrow view of the line in front of the robot. You can also use this viewer to manually adjust the camera focus until the edges of the line are crisp. The following is a screenshot of the PixyMon viewer showing good line detection, with the line in focus and the onboard LEDs turned on:

Figure 12.9 – Live view from the PixyCam, running the line-tracking program

Pixy has two mounting holes, but we cannot mount it directly to the wooden frame of our robot. We need an adapter piece. We could make it out of wood, but this is also an excellent application of 3D printing. Once you know the right mounting position from inspecting the live images with PixyMon, you can take a few measurements with a caliper and use CAD software such as Onshape to create a 3D model of the required adapter piece. If you have a 3D printer, you can directly print it from PLA material.

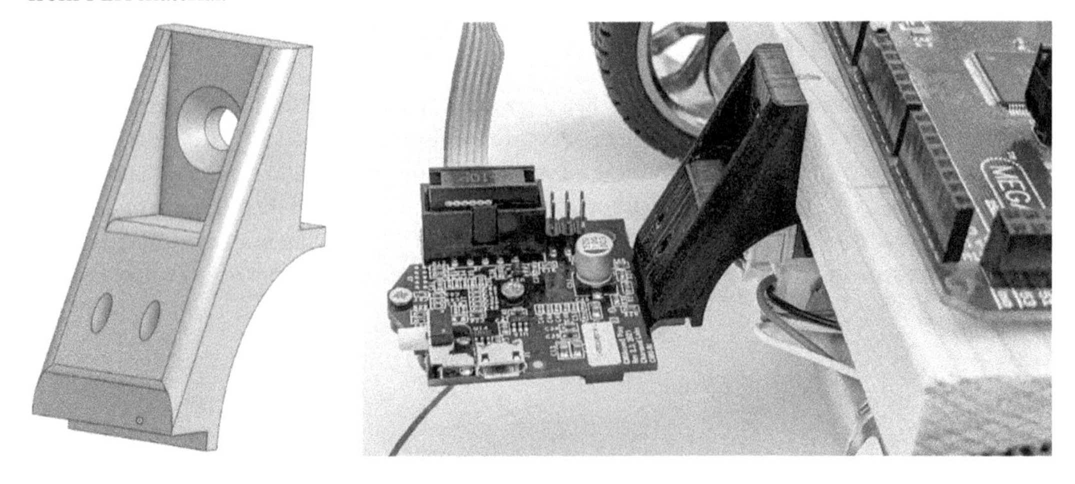

Figure 12.10 – Pixy2.1 mounted to the robot with a custom-designed and 3D-printed adapter piece

It is for pieces like this where the flexibility and speed of CAD, paired with 3D printing, can tremendously accelerate the process of building robots.

# Software

Now that our robot is built and ready for action, it is time to write the Arduino program to bring it to life. With everything that we have covered in the past chapters under our belt, this will be the easy part.

## Overview

Before we dive into writing the code, let us define the desired behavior of our robot and how we would like to control it. When the robot is powered on, it is in **Manual control** mode and you can drive it around via Bluetooth with an emulated control pad on your smartphone. We will use the *Adafruit Bluefruit Connect* app for this purpose, which we already got to know in the previous chapter. If we tap the **2** button, the robot switches to **Line following** mode and autonomously follows a black line on the ground (if there is one). If it loses the line, it should stop. By tapping the **1** button, we can bring the robot back under manual control.

## Motor control

Since we have two motors and certainly do not want to write the same motor control code twice, we implement the motor controller as a class. That way, we can simply create two motor controllers as instances of the same class, with all the benefits of OOP. The following is the definition of the `Motor` class for motor control. Most of this code should look familiar to you based on what we have learned about motor control and OOP so far:

```
class Motor {
public:
  // Constructor.
  Motor(int pwm_pin, int dir_pin_A, int dir_pin_B) {
    pwm_pin_  = pwm_pin;
    dir_pin_A_  = dir_pin_A;
    dir_pin_B_  = dir_pin_B;
  }

  void begin(int max_pwm) {
    pinMode(dir_pin_A_, OUTPUT);
    pinMode(dir_pin_B_, OUTPUT);
    digitalWrite(dir_pin_A_, LOW);
    digitalWrite(dir_pin_B_, LOW);
```

```
    analogWrite(pwm_pin_, 0);
    max_pwm_ = max_pwm;
  }

  void setPwm(int pwm) {
    // Constrain pwm to protect the motor.
    pwm_ = constrain(pwm, -max_pwm_, max_pwm_);
    // Set direction according to sign of pwm parameter.
    bool dir = pwm_ > 0;
    digitalWrite(dir_pin_A_, dir);
    digitalWrite(dir_pin_B_, !dir);
    int command_pwm = abs(pwm);
    analogWrite(pwm_pin_, command_pwm);
  }

  void adjustPwm(int val) {
    setPwm(pwm_ + val);
  }

  int getPwm() {
    return pwm_;
  }

private:
  int pwm_pin_ = 0;
  int dir_pin_A_ = 0;
  int dir_pin_B_ = 0;
  int pwm_ = 0;
  int max_pwm_ = 255;
};
```

You might want to test this class on its own, before adding more components to the code. Define the motor pins and instantiate a `Motor` object outside of the `setup()` function, then initialize it inside the `setup()` function:

```
const int left_motor_pwm_pin = 4;
const int left_motor_dir_pin_A = 6;
```

```
const int left_motor_dir_pin_B = 5;
const int max_pwm = 200;
Motor left_motor(left_motor_pwm_pin, left_motor_dir_pin_A,
left_motor_dir_pin_B);
void setup() {
  left_motor.begin(max_pwm);
}
```

In the `loop()` function, you can call the object's `SetPwm()` function, for example, with the scaled-down analog value of a potentiometer that you can attach to pin `A0` for testing:

```
void loop() {
  int pot = analogRead(A0);
  left_motor.setPwm(map(pot, 0, 1023, -max_pwm, max_pwm);
}
```

This code should allow you to control the left motor of your robot in both directions with the potentiometer. If it works as expected, you can instantiate a second `Motor` object for the right motor and test that it works, too.

## Bluetooth interface

We will use the same Bluetooth interface module and the same technique for parsing the Bluetooth data that we learned in the previous chapter. Once we receive the 10 bytes of a control pad message from the phone app, we determine which button was pressed based on the value of the third byte. If it was `'1'`, we switch the robot to manual control, and if it was `'2'`, we engage the automatic line following. If the robot is in manual mode and the button was one of the four arrows (byte values `'5'` through `'8'`), we adjust the speeds of the two motors to go forward, backward, turn right, or left, accordingly.

To manage the two modes of the robot (manual control and automatic line following), we introduce `enum` object called `Mode` and initialize it to `MANUAL_CONTROL`, so that the robot does not start following a line after power-on without receiving the command to do so:

```
enum Mode {
  MANUAL_CONTROL,
  FOLLOW_LINE
};
Mode mode = MANUAL_CONTROL;
```

You can find the full `ParseBle()` function as part of this chapter's code in the GitHub repository. You should again try this code out on its own. You can replace the function calls inside the BLE parser with simple `Serial.print()` commands to ensure that the parser is reacting to button presses as expected.

## Line following

The line following is the most interesting bit of this robot program since it contains image data processing and communication with the Pixy camera. First, you will need to download the Arduino library for the Pixy2 camera from `https://pixycam.com/downloads-pixy2/`. Place the files in the `libraries` folder and restart your Arduino IDE. Like many Arduino libraries, the Pixy2 library comes with several examples that you can look at to familiarize yourself with its API. At the time of writing, the Pixy2 library was not available through the Arduino IDE's library manager.

To use the library, you need to include it at the top of your sketch, instantiate a Pixy2 object, and initialize it in the `setup()` function:

```
#include <Pixy2.h>
Pixy2 pixy;
void setup() {
  pixy.init();
}
```

If you also instantiate and initialize the two `Motor` objects, the following function implements the line following. Pixy provides the start and end point coordinates of the line vector that it identifies, and you can base the line-following algorithm on this information. We can compute the average deviation of the line from the center of the image and use this as an error input to a proportional controller to adjust the wheel speeds. To make the robot also drive forward, we need to add a forward offset to the resulting PWM value:

```
// Parameters of the line following algorithm.
const int line_follow_pwm_offset = 45;
float kp_line = 3.5;

// Line following algorithm:
void followLine() {
  // Get line detection results from Pixy.
  int res = pixy.line.getMainFeatures();
  if (res <= 0) {
    // No line detected. Stop motors and print error message.
    left_motor.setPwm(0);
```

```
      right_motor.setPwm(0);
      ble.println("No line found.");
      return;
    }
    if (res & LINE_VECTOR) {  // Line detected.
      // Lateral pixel coordinate of the far end of the line
  vector.
      long x_far = (long)pixy.line.vectors->m_x1;
      // Lateral pixel coordinate of the near end of the line
  vector.
      long x_near = (long)pixy.line.vectors->m_x0;
      // Average line deviation from frame center.
      long error = (x_far + x_near - pixy.frameWidth) / 2;
      // Line following controller.
      left_motor.setPwm(line_follow_pwm_offset + kp_line *
  error);
      right_motor.setPwm(line_follow_pwm_offset - kp_line *
  error);
    }
  }
```

You can try out any number of improvements to this algorithm, using the information that the camera provides. The Pixy documentation and code examples of the Pixy2 library can be good places to find inspiration.

## Battery monitoring

The battery voltage divider allows us to monitor the battery with very little code. This is how we do it. First, we implement a short function that turns the analog reading of the voltage (`U_sense`) over the low-side resistor (`R_ls`) together with the value of `R_ls` and the value of the high-side resistor (`R_hs`) into the battery voltage:

```
float getBatteryVoltage() {
  const float R_sense = 4.7;   // kOhm
  const float R_series = 10.0;   // kOhm
  float U_sense = 5.0*(float)analogRead(A0)/1023.0;   // V
  // Compute battery voltage in Volts.
  return U_sense * (R_series + R_sense) / R_sense; // V
}
```

And in the `loop()` function, we call the `MonitorBattery()` function as fast as we can. It compares the sensed battery voltage to a minimum voltage. If the battery voltage is below the minimum threshold, this function puts the robot into manual mode, stops the motors, and prints out an error message to the Bluetooth serial port:

```
void monitorBattery() {
  if (getBatteryVoltage() < min_batt_voltage) {
    // Put into manual mode, stop motors and turn off lamp.
    mode = MANUAL_CONTROL;
    left_motor.setPwm(0);
    right_motor.setPwm(0);
    pixy.setLamp(0, 0);
    ble.println("Error: Battery voltage low.");
  }
}
```

You need to declare a `min_batt_voltage` global variable and initialize it to a value that is a little over the minimum allowed voltage for the battery that you use. This will prevent any battery damage from a deep discharge and other possible issues that can stem from a very low battery voltage.

The Bluetooth parser, the line-following function, and the battery monitoring are all that we need to call inside the program's `loop()` function:

```
void loop() {
  parseBle();
  if (mode == FOLLOW_LINE) {
    followLine();
  }
  monitorBattery();
}
```

If you decide to add a blinker to blink a status LED, you will also call its `blink()` function inside `loop`. Keeping the `loop()` function concise and implementing all the details in encapsulated subroutines is a useful design principle to create programs that are easily readable, both for future you and others that might want to reuse your code.

## Hardware compatibility

As we mentioned earlier, the Arduino Mega is a better choice for this robot than the Arduino Uno. However, the Arduino Uno is sufficient, and we can write the code such that it works with both boards using so-called preprocessor directives. The preprocessor is a text-processing software that reads and modifies the code before it hands it to the compiler. Any line in a C++ program that starts with # is a preprocessor instruction. We can use the preprocessor to check whether an Arduino Uno or Mega was selected as the board, and have it delete the code that is meant for the other board before the code is compiled. That way, we can assign pins differently for the Uno or Mega and use `SoftwareSerial` on the Uno but `HardwareSerial` on the Mega. The following block of code gets compiled when we select the Arduino Uno as a board:

```
#if defined(__AVR_ATmega328P__) // Arduino Uno.
#include <SoftwareSerial.h>
const int ble_RX_pin = 2;
const int ble_TX_pin = 3;

const int left_motor_pwm_pin = 5; // 980 Hz PWM.
const int left_motor_dir_pin_A = 4;
const int left_motor_dir_pin_B = 7;

const int right_motor_pwm_pin = 6; // 980 Hz PWM.
const int right_motor_dir_pin_A = 9;
const int right_motor_dir_pin_B = 8;

// Use SoftwareSerial for Bluetooth.
SoftwareSerial ble(ble_RX_pin, ble_TX_pin);
```

If we instead select the Arduino Mega as the board, the following block of code gets compiled. This lets us select the best pins on a per-board basis and lets us use Hardware Serial when possible. Since all the variable names are the same and we create the same reference called `ble` to either the `SoftwareSerial` or `HardwareSerial` port, the rest of the program will be unaffected by these changes:

```
#elif defined(__AVR_ATmega2560__) // Arduino Mega2560.
const int left_motor_pwm_pin = 11;const int left_motor_dir_
pin_A = 5;
const int left_motor_dir_pin_B = 4;

const int right_motor_pwm_pin = 12;const int right_motor_dir_
pin_A = 9;
const int right_motor_dir_pin_B = 8;

// Use HardwareSerial port 2 for Bluetooth.
```

```
HardwareSerial& ble = Serial2; // ble is a reference to
Serial2.
```

If none of these boards were selected, the preprocessor will abort with a custom error message:

```
#else
#error Invalid board. Use Arduino Uno or Mega2560.
#endif
```

This is a nice example to show the use of preprocessor instructions such as `#if` and `#error`. However, in practice, you should use them very sparingly to avoid unnecessary complexity in your code and avoid confusion for you or anyone else who will read and try to understand your code in the future. The biggest downside of preprocessor logic is that the compiler cannot help you to spot bugs. The compiler can only analyze the results of the preprocessor.

# Tuning and testing

There are a few parameters that we need to tune in order for the robot to function as expected. We can start with the directions in which the wheels turn. Our code assumes that if we pass a positive value to a motor's `SetPwm()` function, the associated wheel spins forward. You can test this out by pressing the *up* arrow in the control pad on your phone and checking whether the wheels turn forward. If they spin the wrong way, you can either swap the order of the way the motor cables are plugged into the motor controller, or you can swap the values of the `motor_dir_pin_A` and `motor_dir_pin_B` variables for that motor.

The value of the `button_increment` variable determines how much the `pmw` value is increased for every button pressed on the control pad. The best value depends on the battery and the motors that you use. One press should be just enough to get the motors spinning reliably. You can do some testing and adjust the value accordingly.

The value of `line_follow_pwm_offset` determines how fast your robot drives when following a line. You can set this just a little higher than `button_increment`. Once you improve the line-following algorithm, you can hopefully let your robot go much faster.

The `max_pwm` variable determines how much *throttle* your robot can use, with 255 being the maximum. Setting this value below 255 is only necessary if you have concerns that the battery voltage is too high for your motors and might damage them if applied fully.

The `kp_line` proportional controller gain determines how much the robot steers toward the center of the line. This is the most fun to play with. A small value will make the robot barely react to the line, while a large value can make it start shaking, losing the line very quickly. If you improve the line-following algorithm, you will most likely end up with one or more additional parameters that you need to tune to achieve the best controller performance. It is usually a good strategy to start with very small values and study the effect as you slowly increase them. With some additions to the code, you can even use the Bluetooth interface to change parameters online, while the robot is running. Similar to how we use the arrow buttons to change the wheel speeds, you could just as well use them to change controller parameters and see the effect right away. If you do that, make sure to remove the `const` keyword in front of them, otherwise, the compiler will not allow you to change the value.

# Summary

In this chapter, we put together a lot of what we learned in this book so far to create a Bluetooth-controlled, camera-equipped, line-following robot from scratch. This robot is an ideal starting point for you to explore the control of mobile robots and to try out different controller ideas. There are some straightforward ways to improve upon the simple line-following algorithm. For example, you can use the direction of the line vector to make line following more robust and allow your robot to go faster around the track. The Pixy camera can also detect intersections of lines and turn signs, and you can experiment with these features. How about using one of the controller buttons to let the robot turn around and continue to follow the line in the opposite direction? Or changing controller parameters on the fly and seeing the effects in real time, without having to re-upload the program? Or plotting the deviation from the line in the live plotter on the phone app?

You can also use this little robot for other purposes. You can give it additional sensors or a different wireless interface to send it on longer missions. You can use the motors' incremental encoder to implement accurate position or velocity control and see how well you can make the robot go in a perfectly straight line or trace a square on the floor. The purpose of this little robot is to let you dive head-first into the world of Arduino robots and try out everything that you have learned so far.

In the next chapter, we will build on much of what we did in this chapter and turn our line follower into an even more advanced radio-controlled, self-balancing, telepresence robot with even more features.

# Further reading

Since line following robots are popular projects among DIY roboticists, you can find a lot of related projects online. Following is a link to one of them, as wells as link to the PixyCam website where you can learn all about its many features that we did not cover in this chapter:

- A simpler line-following robot with a more basic sensor setup. Designs like this are common beginner projects: `https://www.instructables.com/Line-following-Robot-with-Arduino/`.
- Official website of PixyCam: `https://pixycam.com/`.

# 13

# Building a Self-Balancing, Radio-Controlled Telepresence Robot

This chapter is the final project of the book. We will build a self-balancing robot that you can combine with your smartphone or tablet to create a telepresence robot. There is no need for a telepresence robot to be self-balancing, but it is a really fun way to explore some interesting new concepts. This robot is based on the line-following robot that we built in the previous chapter, but we will make necessary hardware upgrades and add a few more exciting features to create an even more capable robot. We will also learn how to use sensor fusion for pitch estimation with an IMU sensor, how to estimate wheel velocity from position encoders, and discover some more interesting applications of CAD and 3D printing.

This example project is the most advanced in this book and gives you the opportunity to see a lot of interesting concepts come together. The chapter is packed with hands-on experiments, and is structured as follows:

- Building the robot
- Adding new software capabilities
- Writing and tuning the control system

Even if you choose not to build this robot in its entirety, you will surely find the new concepts that we will introduce interesting and useful.

## Technical requirements

This chapter builds on the previous chapter by extending the line-following robot. To make the new robot, you will need the following:

- Powerful motors with encoders (`https://www.pololu.com/product/2822`), wheel hubs (`https://www.pololu.com/product/1083`), large wheels (`https://www.pololu.com/product/1435`), and metal mounting brackets (`https://www.pololu.com/product/1084`)
- A six-axis IMU with SPI interface (`https://www.adafruit.com/product/4480`)
- A selfie stick, optionally with single-axis stabilization, such as `https://www.amazon.com/dp/B086ZRLJ92`
- An RC system (transmitter and receiver), such as `https://www.amazon.com/dp/B09BTSJN7P`

## Building the robot

Since our self-balancing robot is based on the line follower, there is not too much to do in terms of building it. However, there are a few modifications we need to make. Most importantly we need to use more powerful motors to enable robust balancing, and we need a phone holder that sticks out far above the top of the robot for the telepresence functionality. There are also two new electrical components, an IMU for tilt estimation and an RC receiver to make it radio-controlled.

To prepare for these additions, you will first need to remove the PixyCam camera, the two driven wheels (if you have not been using the bigger motors yet), and the passive caster wheels from the robot.

### Integrating motors and encoders

This robot needs a little more power to achieve robust balancing than the little motors that we used for the line-following robot can provide. For this robot, we are using 12V motors with a 19:1 gearbox and encoders with 64 **counts per (motor) revolution** (**CPR**), which translates to 1,216 counts per wheel revolution. We need encoders with this much resolution to reliably estimate the wheel velocity as an input to the robot's control system. We will also add larger wheels to give our robot a little more ground clearance and to increase the maximum speed.

Just like in the previous chapter, we will use metal mounting brackets to attach the motors to the wooden crossbar of the robot frame. Only this time, we attach them to the narrow side of it, which is now the lowermost part of the frame. Since the mounting holes in the bracket are spaced too far apart, we cannot simply use screws to fasten the brackets to the wood. But there are at least three practical alternatives that we could consider:

- We could design and 3D-print an adapter piece that holds the motor brackets and attaches to the wooden frame

- We can drill new mounting holes in the center of the motor bracket and use screws to attach the brackets to the robot
- We can use double-sided mounting tape to attach the brackets to the frame and secure the connection with cable ties

The last option is the simplest, so let us select this one for now (feel free to make another choice for your version of the robot though!). With a bit of double-sided mounting tape and two sturdy cable ties, we can quickly mount the motors in a way that is at least *good enough* for a prototype.

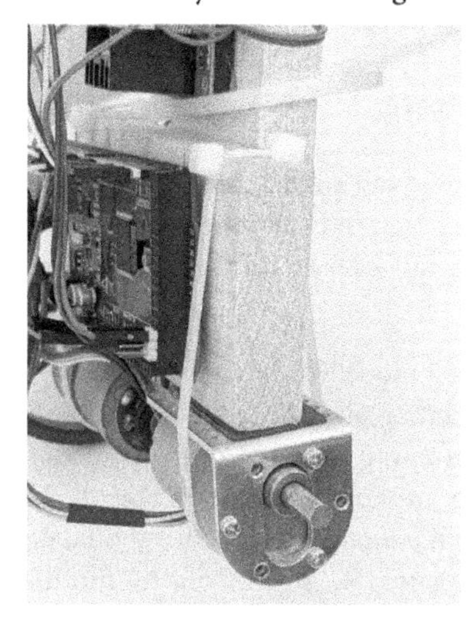

Figure 13.1 – One of the motors, mounted with a mounting bracket, double-sided tape, and secured with cable ties

The motors and encoders also need to be connected electrically. You can connect the two poles of each motor (red and black) to the outputs of the motor driver board, just as we did with the smaller motors on the line follower. The encoders have four additional wires: **5V**, **GND**, **A**, and **B**. Connect 5V (blue) and GND (green) to the two LV power rails on the solderless breadboard, and the two signals A and B (yellow and white) to two Arduino pins that are interrupt capable. The following code snippet shows the pin numbers:

```
namespace left {
const int kEncoderPinA = 20;
const int kEncoderPinB = 21;
}
namespace right {
const int kEncoderPinA = 18;
const int kEncoderPinB = 19;
}
```

The `namespace` keyword is a feature of the C++ programming language that we will get to know in the software section of this chapter. You can see that the encoder signals of the left encoder are connected to pins 20 and 21, whereas the signals of the right encoder are connected to pins 18 and 19. These four pins are all hardware interrupt capable on the Arduino Mega. It might be surprising to you that we use the same variable names, for example, `kEncoderPinA`, twice in this code snippet. Should variable names not be unique? This is possible because all variable names are unique within a namespace (or scope). If we want to use the variable names outside their namespace, we need to prefix them with the namespace name and a double colon, and thus `left::kEncoderPinA` and `right::kEncoderPinA` are unique again.

## Integrating the RC receiver

Next, we need to add an RC receiver to our robot. Using a hobby remote control makes controlling the robot very intuitive, especially if you have experience with remote-controlled cars or planes. While it limits the control to closer ranges than cellular data for example, its simplicity makes it a great starting point for this project.

The RC receiver, no matter what exact model you have, is extremely easy to integrate. You can simply stick it to the frame of your robot with a piece of double-sided mounting tape. For the best signal quality, you will want the antenna(s) to point up. Connect any of the GND and 5V pins of your receiver to the corresponding power rails of the solderless breadboard. RC receivers draw very little power, and the voltage regulator on the motor driver board should not have any problems powering it. Our Arduino needs to decode only two RC signals: one for throttle and one for steering. For that to work, all that is left to do is connecting the signal pins of the two channels you want to use to two interrupt-capable pins of the Arduino. On the Arduino Mega, we have pins 2 and 3 left for this, and we use these pin numbers in the following code snippet:

```
namespace rc {
namespace steer {
const int pwm_input_pin = 2; // Receiver channel 1.
}
namespace throttle {
const int pwm_input_pin = 3; // Receiver channel 3.
}
}
```

This is an example of a **nested namespace**. Outside the namespace called `rc`, we can access these two variables as `rc::steer::pwm_input_pin` and `rc::throttle::pwm_input_pin`. The outer namespace is only here to demonstrate the concept of nesting and is not actually necessary. You will not find it in the actual program code on GitHub.

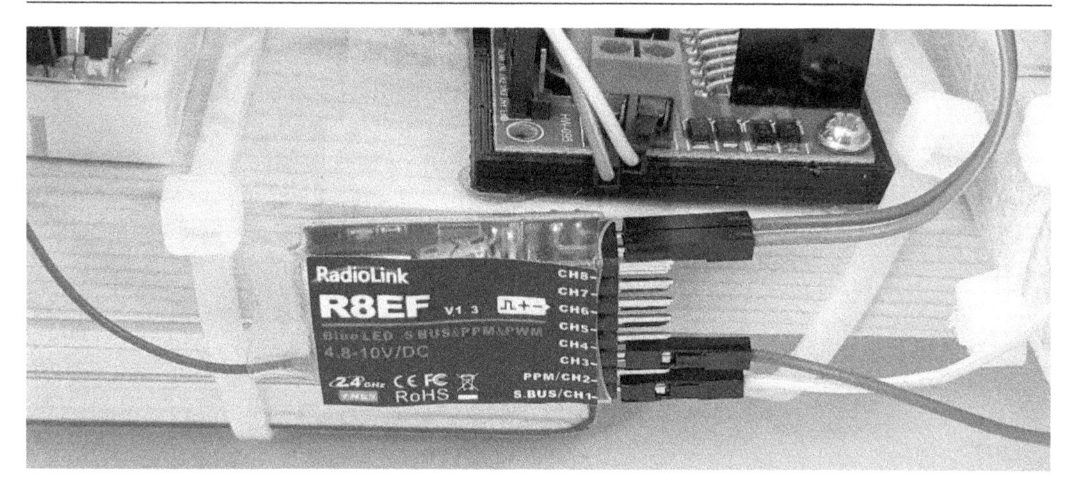


Figure 13.2 – The RC receiver is attached with double-sided tape. Two wires are used to power it, the other two are signal wires going to the Arduino

As an aside, many modern RC receivers support not only PWM outputs but also **Pulse Position Modulation** (**PPM**) outputs. With PPM, many channels can be encoded in a single electrical signal. This makes it possible for your Arduino to read all channels of the receiver with just a single interrupt input, as opposed to needing one input per channel for PWM. However, decoding PPM can be a significant load for a slow Arduino such as the Mega2560 and interfere with other functions. Therefore, using PWM is the better choice for our telepresence robot. However, if you decide to use a more powerful board such as a Teensy 3.5, using PPM is the better interface if your receiver supports it. You can find code for PPM decoding in this chapter's folder on GitHub.

## Integrating the IMU

Perhaps the most important hardware addition to this robot, the one that allows it to balance, is the IMU. We need it to generate the input signals for the balance control algorithm. Specifically, we will use it to estimate the pitch of the robot (this is what we want to control) and to measure the angular rate at which it is pitching (this is the time derivative of what we want to control). To measure the angular rate, all we need is a single-axis gyroscope aligned with the pitch axis of the robot (the spatial axis connecting the wheel centers). Unfortunately, there is no sensor that allows us to sense the tilt angle directly, so we must estimate it. For the estimation algorithm, which we will explore a little later, we need the same gyroscope and a two-axis accelerometer in addition. Sensors with a three-axis accelerometer and a three-axis gyroscope are quite common components, often called six-axis IMUs. A six-axis IMU has all the sensing capabilities we need and more. An additional requirement is a fast SPI interface that allows us to sample the sensor several hundred times per second to get good estimation accuracy. ST's *LSM6DS33* is a great sensor for our purposes, and Adafruit makes a nice breakout board for it (`https://www.adafruit.com/product/4480`). Adafruit also provides the `Adafruit_LSM6DS33` Arduino library for this sensor, which you can add to your Arduino IDE via the library manager, making it amazingly easy for us to use this sensor in our robot.

The Adafruit breakout board for the LSM6DS33 can be mounted on the solderless breadboard on our robot once you solder on the pin headers. This makes even the hardware integration into our robot quite easy. The board has a little coordinate system (arrows denoted by X, Y, and Z) printed onto it. You need to mount it such that the Y arrow points up, away from the wheels. Otherwise, you will need to make some adjustments to the code to account for the different orientations of your sensor. Once on the breadboard, you can connect it to 5V, to GND, and to the SPI pins of the Arduino:

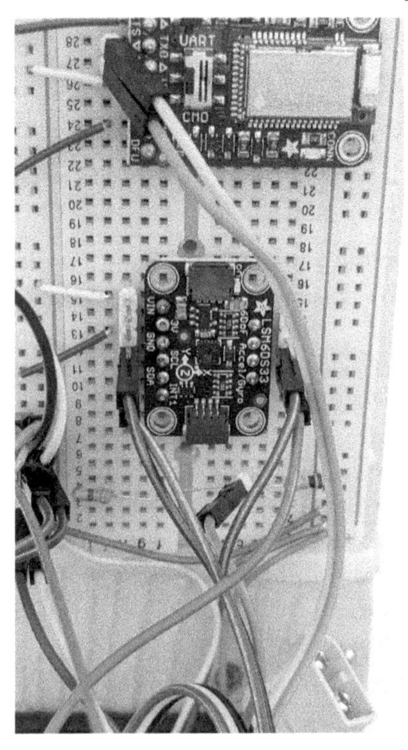


Figure 13.3 – The IMU, mounted on the robot's breadboard. Note the little arrow marked Y pointing up

Adafruit even has a guide to help you get started (`https://learn.adafruit.com/lsm6ds33-6-dof-imu=accelerometer-gyro`). On the Arduino Mega, the SPI pins are as follows:

```
// MOSI: 51 --> LSM6D33 SDA
// MISO: 50 --> LSM6D33 DO
// SCK: 52 --> LSM6D33 SCL
```

We also need to choose which pin to use for the chip select signal:

```
const int kSpiCsPin = 10; // --> LSM6D33 CS
```

You can use any available pin; `10` is just an arbitrary choice.

## Phone holder

What makes the robot we are building here a telepresence robot is its ability to carry around a phone or a tablet to run a video conferencing app such as FaceTime, Google Meet, Skype, or Zoom. You can sit in front of a screen at the other end of the line and use the robot to be virtually present where the robot is. To get a pleasant experience out of this, we need the phone to be high above the floor. That means the phone holder should extend up quite a bit. Since the robot will pitch forward and backward a little as it drives around and balances on its two wheels, the phone holder should also be able to compensate for this motion to provide us with a stable camera image. There are several ways in which we can get to a stabilized phone holder. We can, of course, just build one ourselves, using an RC servo motor to adjust the angle of the phone to compensate for the changing pitch of the robot, or we can use an off-the-shelf product. In fact, there are selfie sticks that are actively stabilized around one axis that are almost exactly what we need. For our robot, we are going to go down this route and use a low-cost selfie stick as a phone holder (`https://www.amazon.com/dp/B086ZRLJ92`). There is one caveat though: the selfie stick stabilizes the phone around the wrong axis. So, we need to design and 3D-print two adapter pieces that let us mount the phone on the selfie stick with a rotational offset of 90 degrees.

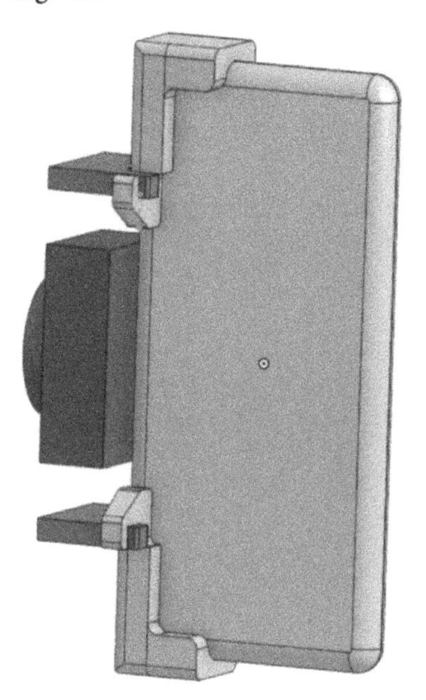

Figure 13.4 – CAD model of the phone holder, and the real assembly with printed adapter pieces

Since selfie sticks are meant to be hand-held and not robot-mounted, we need to find a way to attach ours securely to the wooden frame of the robot. We can again resort to CAD and 3D printing for this, designing and printing an adapter that securely connects to a robot on one side and has a hole that snugly fits the handle of the selfie stick on the other.

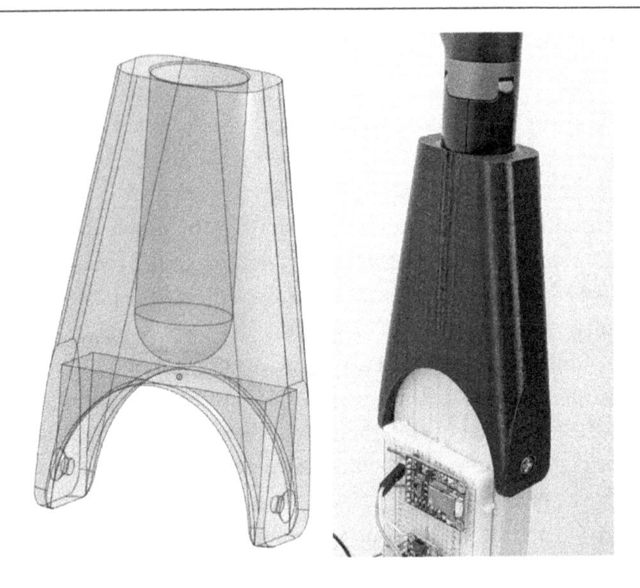

Figure 13.5 – CAD model of the selfie-stick adapter and the 3D-printed version mounted to the robot, secured with two screws at the bottom

And with these hardware changes out of the way, here is the final robot, ready for testing:

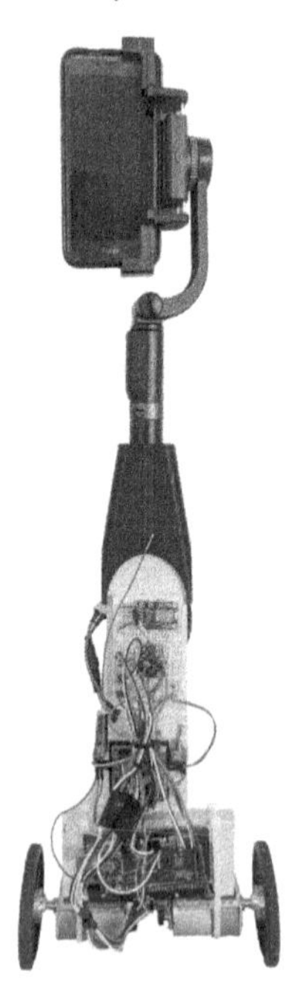

Figure 13.6 – The completed robot for this chapter (with the phone mast retracted), ready for testing

Let us now dive into the Arduino code for your new robot. We have a lot to cover!

# Writing the control software

The Arduino program for this robot is more extensive (in terms of lines of code) than all the other ones we developed so far. We will not print the entire code here, but only sections of it that will help you understand the key concepts. You can always refer to the full code in the GitHub repository that accompanies this book.

## Driving the motors

We are using the same motor driver board that we used in the previous chapter, so we could just reuse the motor driver code without changes. However, simply copying code from one project to another is not the best way to do this, since older projects will not be able to benefit from improvements you make in newer projects. The solution to this problem, you guessed it, is to package the motor driver code as a library!

### *Motor driver library*

We already learned how to turn object-oriented code that you want to reuse into a library. Our motor driver library consists of a `MotorDriver` class that exposes an interface that lets you control one motor. For a robot with two motors, we can simply instantiate two `MotorDriver` objects, without having to copy any code at all. Instead of looking at the entire code, let us just take a look at the public functions (the library's API) defined in the header file:

```
// Constructor.
MotorDriver(int pwm_pin, int dir_pin_A, int dir_pin_B);
// Initialization. Call inside setup().
void begin();
// Set PWM value. Negative values reverse direction.
void setPwm(int pwm);
// Set the maximum admissible PWM value.
void setMaxPwm(int max_pwm);
// Change PWM value by a certain value.
void adjustPwm(int val);
// Get the current PWM value.
int getPwm();
```

The first function is the constructor, which takes three pin numbers, one for the PWM pin and one for each direction pin. The second function, `begin()`, initializes the hardware with calls to `PinMode()`. It is good practice to collect function calls that initialize hardware in a dedicated API function and have the user call it inside the program's `setup()` function. This is more reliable than initializing hardware in the constructor. The style guide for Arduino libraries encourages calling this API function `begin()`, which is why you see many libraries use that name for it (`https://docs.arduino.cc/learn/contributions/arduino-library-style-guide`). These and all the other API functions will look familiar to you from the `Motor` class of the previous chapter.

### *Adjusting the PWM frequency*

There is one more tweak that we make to how we drive the motors. When we learned about driving DC motors, we briefly mentioned that the frequency of the PWM signal has two effects: it influences the motors' torque and the noise that the motors make. Our line follower from the previous chapter uses the Arduino's default PWM frequency and, therefore, makes a humming noise as it drives around, which you surely noticed. We do not want our telepresence robot to make the same humming noise, so we need to push the PWM frequency above 20 kHz, outside the audible range. There is no built-in Arduino function to do this, so we need to do this the old-school way, by studying the microcontroller's datasheet and changing the content of control registers, bit by bit. What results from this are the following two lines of code that raise the PWM frequency for the motor to around 31 kHz:

```
// Change PWM frequency on pins 11 and 12 to ~31.4 kHz.

// Timer/Counter1 prescaler setting for motor PWM.
const int kTimer1PrescalerSetting = 0b001;
TCCR1B = TCCR1B & 0b11111000 | kTimer1PrescalerSetting;
```

These lines look cryptic, so let us understand what they mean by peeking a little bit into the inner workings of your Arduino Mega. The ATmega microcontroller of your Arduino Mega uses a piece of hardware called a **Timer/Counter** to generate PWM. It has, in fact, three Timer/Counters that all generate PWM slightly differently. Timer/Counter 0 produces the PWM on pins 4 and 13, but it is also used for important functions, such as `millis()` and `micros()`, so we'd rather not change any of its settings to not break things. Timer/Counter 1 produces the PWM for pins 11 and 12 and is not used for anything else, so let us try to change its PWM frequency. This timer produces PWM in **phase-correct mode**. That means it first counts from 0 to 255, then down to 0 again, and repeats this process indefinitely. It turns the PWM output off (signal low) when it hits the **compare value** (the PWM value we set, for example, `200`) on the way up, and turns the output on (signal high) again when it hits the same value on the way down. This means it produces one PWM period for every cycle of counting up and down, or every 510 counts. The timer counts at the same speed as the clock cycle of the CPU (16 MHz), divided by a prescale factor or **prescaler**. This means that in order to change the PWM frequency, we need to change the prescale factor. The Arduino default prescaler for Timer/Counter 1 is `64`, which means it counts at a frequency of 250 kHz (16 MHz / 64) and therefore, produces a 490 Hz PWM signal (250 kHz / 510 steps per PWM period). If we instead set the prescaler to `1`, it generates a 31.4 kHz PWM frequency (16 Mhz / 510), which is well outside the audible spectrum and thus leads to quiet motors. This explains the first line of the code snippet: we want the prescaler for timer 1 to be set to `1`. And because we are going to use this value in bit manipulation, we write it as a binary value.

> **Note**
>
> This setting is not always the same as the resulting prescaler value, but for a prescaler of `1`, the setting is also `1`. You can find the full mapping from the prescaler setting to the prescaler value in the microcontroller datasheet.

But what do we do with this setting? Timers/Counters are controlled by **Timer/Counter Control Registers**, and you can find the meaning of all the bits in these registers in the microcontroller's datasheet. The prescaler setting needs to be written to the three lowest bits of the Timer Counter Control Register for Timer 1, index B, called *TCCR1B*. The second line of code is what changes this register. First, we compute the binary AND between the current content of this register and the bitmask `0b11111000`. This leaves the upper 5 bits unchanged thanks to the 1s in the bitmask and clears the lower 3 bits. We then perform a binary OR between the result and the setting we want to write, which applies our setting to the three lower bits of TCCR1B.

This is a good example to highlight the meaning of Arduino. Without a convenient Arduino function, working with microcontrollers is difficult and tedious, and involves carefully studying extensive datasheets and modifying individual bits in cryptically named control registers. In contrast, Arduino functions make much of the functionality of microcontrollers accessible in an easy and intuitive way. But it is also a reminder that Arduino lets you only do so much, and if you want to use every bit of performance or access all functionality of your microcontroller, you need to go beyond Arduino, dive into the datasheet, and start manipulating individual control bits.

## Encoder library

You already know how to use hardware interrupts to read the signals from quadrature encoders like the ones on the wheels of our robot. To make sure that we do not have to duplicate any code to serve our two encoders, we will also put the encoder code into its own library. Unfortunately, we need to call the `attachInterrupt()` function in the main program and cannot call it inside the library, which means this encoder library is not a great example of object-oriented code. Instead, our `QuadEncoder` library exposes two public functions, `changeIsrA()` and `changeIsrB()`. After instantiating a `QuadEncoder` object, we need to wrap its two ISRs in global functions that we can then pass to `attachInterrupt()` in the `setup()` function of the main program. The following snippet demonstrates all of that:

```
#include "QuadEncoder.h" QuadEncoder encoder(4, 5); // Encoder
connected to pins 4 and 5.
// Global wrappers for the encoder's interrupt handlers.
void IsrA() {
  encoder.changeIsrA();
}
void IsrB() {
```

```
    encoder.changeIsrB();
  }
  // Attach encoder interrupt handlers to interrupt sources.
  attachInterrupt(digitalPinToInterrupt(4), IsrA, CHANGE);
  attachInterrupt(digitalPinToInterrupt(5), IsrB, CHANGE);
```

Having to create global wrappers for the interrupt handlers is another little sacrifice we need to make to get the convenience of the Arduino ecosystem.

## Wheel velocity estimation

Encoders only measure the wheel position, but for the control system of our self-balancing robot, we also need the wheel velocity. Velocity is just the change of position divided by time, so at first glance, it seems easy enough to compute the velocity from the position. Let us consider the two most straightforward approaches and why they are not sufficient for our needs.

In **Approach 1**, we measure the **time** (T) between two consecutive encoder ticks. Then, 1/T is the velocity estimate.

This is very intuitive. However, the problems start when the wheel turns very slowly. We only get a new velocity update when an encoder tick happens, which can take much longer than the loop rate of our control system. This means our controller runs on stale velocity data, which can lead to stability problems. With this method, the velocity estimate also never gets to zero, even if the wheel is completely still for a long time. Further, if the encoder is not perfect (and no encoder is), consecutive ticks will not happen at equal time intervals, even if the wheel spins with perfectly constant velocity, leading to a noisy velocity estimate.

In **Approach 2**, we sample the encoder value at fixed time intervals. The position change between samples divided by the time interval is the velocity estimate.

Again, this method is intuitive, and it gives us new velocity estimates at the rate we need. However, in practice, the following tends to happen: if we chose a time interval short enough such that we get high-frequency updates, the encoder might have only incremented one or two positions, even if the wheel spins fast. This is due to the low resolution of hobby-grade motor encoders and means that this method produces a very low-resolution velocity measurement – too low to be useful in our control system.

A better albeit more complicated approach is to use the method from Approach 2 but to keep the latest *n* velocity estimates in a **sliding window** and compute the velocity as the average of the velocities in this sliding window. This still gives us high-frequency updates as the sliding window content changes with every sample, but it also gives us much better resolution as the sliding window contains measurements taken over a longer period. This method, called **moving average**, also has the benefit of filtering out high-frequency measurement noise, at the expense of introducing a bit of lag. *Figure 13.7* shows the raw and filtered velocity estimates of Approach 2 when one of the robot wheels is spun back and forth by hand:

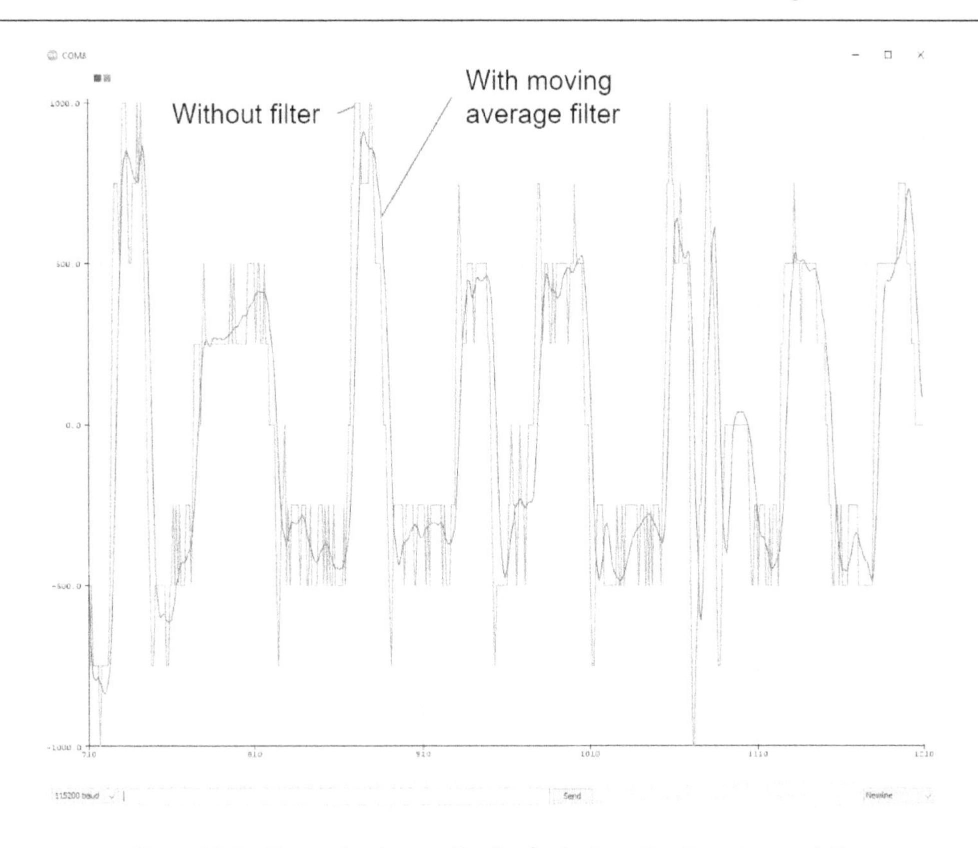


Figure 13.7 – Comparing two methods of velocity estimation: Approach 2 only (noisy, low resolution) and Approach 2 with moving average (smooth with a small lag). This plot was generated with the Arduino IDE 1.8

All these characteristics can be tuned with two parameters, the sampling interval and the size of the sliding window (the number of samples). In our `QuadEncoder` library, the sliding window is implemented as an efficient **ring buffer**. This is a useful concept to know, so make sure to take a look at the library code on GitHub. In our main program, all we need to do is to call the `update()` function of the encoder objects inside the `loop()` function to always have access to an up-to-date velocity estimate via the `getVelocity()` function. The library API also contains the `getRawVelocity()` function, which gives you the unfiltered result of Approach 2. You can use this to plot both and study the effects of changing the sampling rate and the size of the sliding window.

## Namespaces

Because our robot has a left and a right wheel and thus a left and a right motor and a left and a right encoder, there is a lot of code that is identical except for the word *left* or *right*. For example, we need a left and a right version of each encoder pin, each PWM pin, each `MotorDriver` object, and so on. One way to deal with this is to give all the affected variables a `_left` or `_right` suffix. However, this can make variable names very long and is also prone to copy and paste errors. Another, often better way to deal with this situation is to use identical code for left and right but put them into different **namespaces**. This reduces the places in which programming errors can happen and helps make the structure of the code easier to understand. As we saw earlier, variables and functions in different namespaces can have the same name without producing errors, and outside their namespaces, we can use them by prefixing them with their namespace name and a double colon.

Namespaces are one more tool in your toolbox when writing Arduino programs that can help you keep your code organized and free of bugs. Sometimes namespaces are great, but sometimes it is better to create a new class rather than using namespaces. For example, we could create a class called `RobotSide` that holds all variables and functions that are specific to one side of the robot, as well as the API to use them. We could then instantiate two objects of the `RobotSide` type: `robotside_left` and `robotside_right`. As you write more programs, gain more experience, and develop your own style, you will be able to decide which tool is the right one to keep your code clear and error-free in any specific situation.

## Pitch estimation with the IMU

As we mentioned, when selecting the IMU, the balance control system of the robot needs the pitch angle as an input, but there is no way of measuring it directly with an IMU. We need to estimate it from the sensors we have, just like we need to estimate the wheel velocity from the motor encoder. To understand the final estimation algorithm, let us first look at two simple but insufficient ways we could do this estimation.

In **Approach 1**, we integrate the pitch angular rate over time. This integral is the pitch angle.

The idea of this approach is that while we cannot measure the pitch angle directly, we can measure its derivative with respect to time, the angular rate around the pitch axis. This is what the gyroscope of the IMU measures. So, in theory, all we need to do is to integrate the pitch rate with respect to time, and out comes the pitch angle. Writing the code for this is very simple, and it leads to an exceptionally smooth estimate thanks to the properties of the integral to filter out measurement noise. Unfortunately, this approach has a severe problem in practice. Any offset in the rate measurement is exacerbated over time by the integration, and the pitch estimation very quickly drifts far away from the true tilt angle. And with small, hobby-grade MEMS IMUs, there is always an offset that is extremely difficult to compensate for. There is also a systematic numerical error introduced by the digital integration in finite time intervals that our Arduino needs to perform. While this approach can give exceptionally good tilt estimates for a brief period, the result starts to drift away in as little as a few seconds. Using this method alone is only feasible with very high-grade gyroscopes, the kind you would find in submarines or weapons systems.

In **Approach 2**, we use two accelerometers, both perpendicular to one another and perpendicular to the pitch axis. The two-argument arctangent of their two values is the pitch angle.

The idea behind this approach is that every accelerometer, when it stands still, senses the gravitational acceleration caused by earth's gravity, 9.81 m/s^2, or 1g. How much of that acceleration it senses depends on the accelerometer's orientation. If it is perfectly aligned with gravity, it senses the full

1g. If it is perpendicular to gravity (parallel to the floor), it senses none of it. Now, if we have two accelerometers perpendicular to one another, we can look at both their values to see how much and in which direction the robot is tilted. If the first measures 1g and the second 0g, the robot is tilted 90 degrees one way. If the first measures 0g and the second 1g, the robot is tilted 90 degrees the other way. And if both sense the same gravity acceleration, the robot is perfectly upright. The math comes out such that the exact tilt angle of the robot can be computed as the two-argument arctangent (`atan2`) of both values. The advantage of this method is that it does not suffer from any drift, since we do not need to integrate anything, and we always have gravity as a steady reference. The big downside, however, is the fact that this computation will only lead to the pitch angle when the robot is standing still. If it is moving and experiencing dynamic accelerations caused by its motion in addition to gravity, the pitch estimation will get very noisy.

You might have noticed that these two methods have complementary strengths and weaknesses. Approach 1, using the gyroscope, delivers a smooth pitch estimate that is great over a short time horizon, but suffers from drift in the long run. Approach 2, using the accelerometers, never drifts, even in the long run, but is plagued by noise caused by the robot's motion. To get a truly useful tilt angle measurement, we need to combine the strengths of both approaches to overcome their individual weaknesses. Problems like this, where different sensing capabilities are good only in certain ways or under certain conditions and one needs to combine the capabilities to get good results, are quite common in robotics. The solution is called **sensor fusion**, and there are several algorithms (called **filters**) for it. We will use a simple one for our problem, the **complementary filter**. In this algorithm, we can do most of the estimation with the gyro integration to get a precise and low-noise estimate, but we throw a little bit of the accelerometer estimation in there to avoid the result from drifting away. Let us look at the following code snippet to understand how it works. This example is a little simplified, but you can find the full implementation in the `CompFilter` library on GitHub:

```
void update(float rate_z, float accel_x, float accel_y, float
sampling_interval) {
  // Update the global  angle_filtered_ variable via a
complimentary filter.
  // Pitch estimate from accelerometer data.
  float angle_accel_ = atan2(accel_x, accel_y);    // Tilt
change estimated from gyroscope data.
  float gyro_delta = rate_z * sampling_interval;   float
filter_coeff = 0.99; // Weighing of the integral term.
// Complementary filter update.
  angle_filtered_ = filter_coeff * (angle_filtered_ + gyro_
delta) + (1.0 - filter_coeff) * angle_accel_;
}
```

This function first computes an angle estimate based on Approach 2, using the `atan2()` function of the accelerometer readings in the *x* and *y* direction, which are perpendicular to the pitch axis. It then computes the pitch change by multiplying the current angular rate around the pitch axis (the *z* axis) by the sampling interval. This is part of Approach 1. And in the last line, it does the filter update, combining Approach 1 and Approach 2. For Approach 1, the gyro integration, this line would simply be as follows:

```
angle_filtered_ = angle_filtered_ + gyro_delta;
```

For Approach 2, only using the accelerometer data, the last line would be as follows:

```
angle_filtered_ =  angle_accel_;
```

But instead, we take the weighted average of the two, weighing the term of Approach 1 with `filter_coeff` and that of Approach 2 with `1 - filter_coeff`, ensuring that the two weighting factors add up to `1`. This clever line ensures that the gyroscope integral is biased towards the accelerometer estimate in the long term, but without losing its fidelity or inheriting the noise coming from the accelerometers. We can tune the filter performance to our needs by changing the sampling interval (shorter is better) and the filter coefficient. The more drift our gyroscope estimate has, the lower we want to set this value to compensate for it. The following figure shows the three methods in comparison, and you can reproduce them with the `CompFilter` library on GitHub.

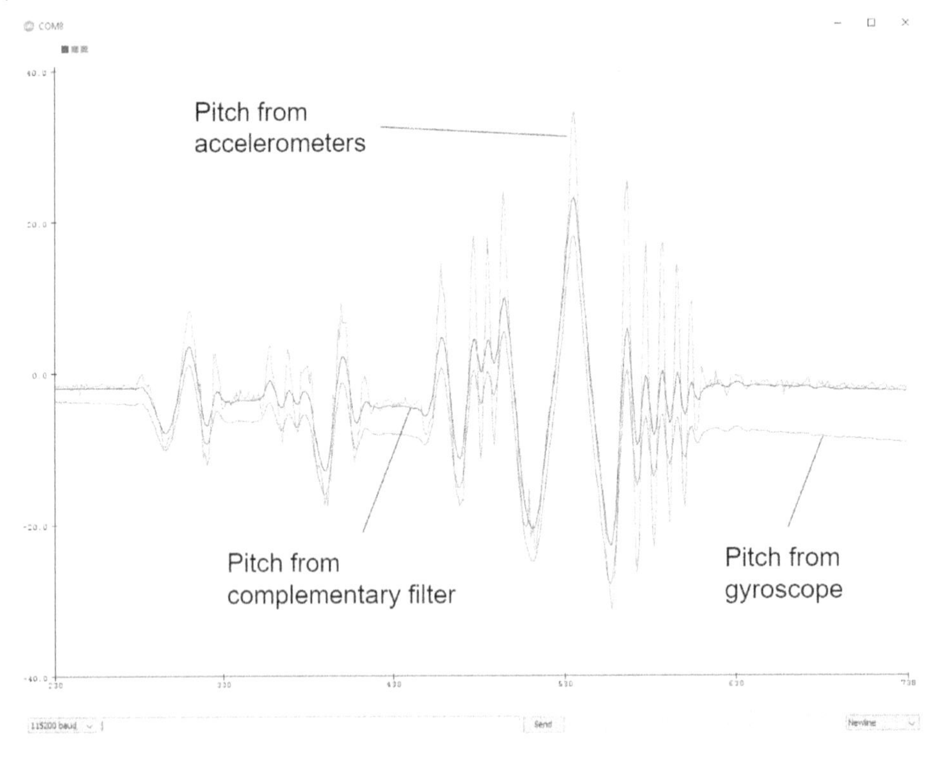


Figure 13.8 – Comparing the three ways of estimating the robot pitch. This plot was generated with the Arduino IDE 1.8

In the code for this robot, we perform the filter update on the `CompFilter` object called `pitchFilter` right after we sample the IMU, feeding in the latest IMU measurements:

```
// Sample IMU.
imu.getEvent(&accel, &gyro, &temp);
pitchFilter.update(gyro.gyro.x, accel.acceleration.z, accel.
acceleration.y, imu_task::kInterval * 1000);
```

Note that, in the way we mounted the IMU, the pitch axis is `x` and the two relevant accelerometer axes are `y` and `z`, respectively. The scaling factor of `1000` converts the `kInterval` variable of the `imu_task` from milliseconds to microseconds to match the CompFilter API.

Since this approach still relies on integrating the gyroscope rate, it is beneficial to remove any systematic offset from this measurement that is inevitable in small MEMS IMUs. This is why the `CompFilter` library allows us to set an offset via `setGyroOffset()`. To measure this offset value (`gyro_x_offset`) after reset, we compute it as the average over several consecutive samples in the `setup()` function:

```
float gyro_x_offset = 0.0;
for (int i = 0; i < kNumImuCalibSamples; i++) {
  imu.getEvent(&accel, &gyro, &temp);
  gyro_x_offset += gyro.gyro.x;
}
gyro_x_offset /= (float)kNumImuCalibSamples;
pitchFilter.setGyroOffset(gyro_x_offset);
```

You just need to be sure to keep the robot still (lying on the ground) during this brief calibration to measure the true offset rather than the actual motion.

The `CompFilter` library provides the filtered angle estimation via its API function, `getAngleFiltered()`, and this is what we use in the robot control system. But it also provides the other two estimation methods via `getAngleAccel()` and `getAngleGyro()` for you to experiment with (for example, to recreate the plot shown in *Figure 13.8*) and to build your understanding of how the complementary filter works.

## Controlling the robot

With all that we have discussed so far, we have the inputs and outputs of the control system in place. The inputs are as follows:

- The sensed PWM pulse widths from the RC receiver
- The motor encoder velocities estimated with the moving average

- The pitch rate sensed by IMU's gyroscope
- The pitch angle, estimated from the IMU's gyroscope, and the accelerometers by the complementary filter

The outputs of the control system are the motor PWM values for the left and right motor drivers. The goal of the control system is to drive the motors such that the robot 1) automatically balances upright and 2) follows the joystick commands, driving forward and backward, steering left and right, and turning in place. We can, therefore, naturally split the discussion of the control system into two parts: the balance control and the remote control. We can formulate both parts of the control system as individual controllers that both generate motor PWM values for the left and right motor. In the end, we can simply add their PWM outputs and send them to the motor controllers.

More precisely, the controllers each contribute to the value of two variables: `pwm` and `diff_pwm`. The value of `pwm` gets applied to both motors in the same direction and is responsible for balancing and going forward and backward. The value of `diff_pwm` gets applied to both motors in opposite directions (differentially) and is responsible for control around the vertical axis as well as steering. At the beginning of each control cycle, we reset `pwm` and `diff_pwm` to `0` and then compute and add the contributions of all controllers to it. After that, the values of the right and left motor PWM values are computed from `pwm` and `diff_pwm` as follows:

```
left_pwm  = pwm + diff_pwm;
right_pwm = pwm - diff_pwm;
```

The following figure gives you a simplified overview of the control system and how its inputs are mapped to its outputs. **M_L** and **M_R** stand for motor right and motor left, respectively.

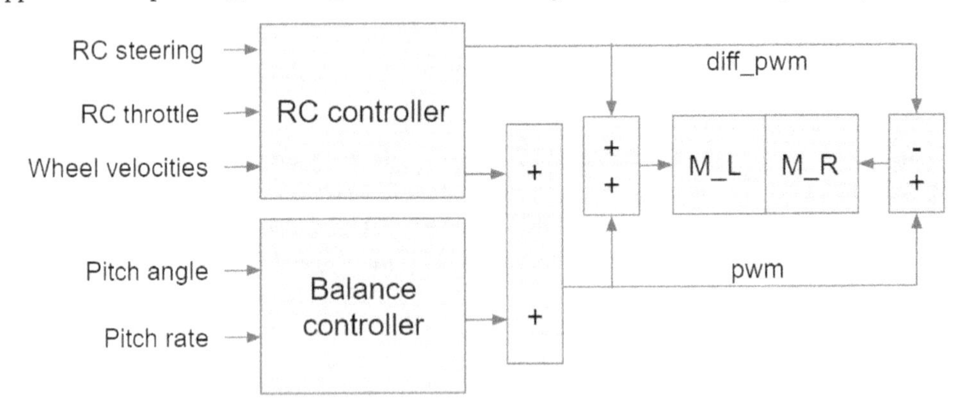


Figure 13.9 – Simplified block diagram of the control system. The feedback paths and filters are omitted for conciseness

Let us look at each controller in more detail now.

## Balance control

If this is your first time building a two-wheeled self-balancing robot, this is probably the part of this chapter you are most curious about. How does the robot keep its balance? The answer is surprisingly simple: a PD controller is all it takes! The difficult part is to estimate the tilt angle, but we have already done that with our complementary filter. The balance controller uses both motors in the same direction, so it only contributes to pwm. The following two lines of code represent the P and the D controller term of the balance controller. Keep in mind that pwm is reset to 0 at the beginning of each control cycle:

```
int pwm = 0;
int pwm = 0;
// P-control.
pwm += kp_balance * (kPitchOffset - pitchFilter.get_angle_
filtered());
// D-control.
pwm += kd_balance * gyro.gyro.x;
```

As you would expect, the P-controller output is simply proportional to the estimated pitch angle, and the D-controller output is proportional to the rotational rate around the pitch axis. We need to find suitable values for the controller gains `kp_balance` and `kd_balance` during controller tuning. But there is one term in the first line that we have not yet talked about, `kPitchOffset`. This constant describes the angle measured by the IMU at which our robot is naturally balancing. This can be 0, but depending on the IMU placement, the weight of the battery and so on it is most likely different from 0. We need to measure this value during controller tuning to make sure our balance controller stabilizes the robot around this equilibrium. If this value is set incorrectly, the balance controller will try to lean the robot forward or backward instead of keeping it upright, causing it to drift away.

## Remote control

This part of the control system is in charge of enabling you to remotely control the robot like a radio-controlled car. We can divide it into two subparts, *throttle* and *steering* control. Throttle control is applied to both motors in the same direction, so it contributes to the value of the pwm variable. Making the robot react to the throttle input is as easy as increasing pwm proportional to the throttle signal:

```
// Convert throttle input to PWM.
pwm += kp_throttle * (float)(throttle::pulse_width_us -
kPwmCenter);
```

The value of the `kp_throttle` gain is something that we need to find during controller tuning. The higher this value is, the quicker the robot reacts to control input, and a lower value leads to smoother control. The current pulse width of the throttle channel is stored in `throttle::pulse_width_us`. But this is not the signal we need. Remember that this value will range from 1,000 to 2,000, depending on the joystick position. To make our calculations easy, we need a value that is centered around 0, and thus we subtract the `kPwmCenter` offset. This is set to `1500`, such that the value in parentheses now ranges from -500 to 500, centered around 0. We further explicitly convert this value to a floating point number with the `(float)` expression to make sure we get the expected result when we multiply it with the gain, which is also of the `float` datatype.

If you tried to drive the robot around like this, you would quickly notice that it tends to drift away forward or backward. This is because there is nothing holding it in place (there is no position controller). To make it a little more stable and easier to drive, we can add a controller term that slows the robot down or, in other words, a D-controller on the robot's position. The derivative of the robot's position is its velocity, so we need the velocity as the input to this controller. Since we are dealing with two wheels, we estimate the robot's linear velocity (as opposed to its rotational velocity around the vertical axis) as the average of the two wheel velocities. Adding a `D-control` term to the position is then as simple as changing `pwm` proportionally to the linear velocity. The sign of `kd_position` ensures that this change happens in the right direction and depends on the specific wiring of your robot:

```
// Compute forward/backward velocity as the average of the two
wheel velocities.
float linear_velocity = (float)(right::encoder.getVelocity() +
left::encoder.getVelocity()) / 2.0;
// Position D-control.
pwm += kd_position * linear_velocity;
```

As usual, we will need to find a suitable value for `kd_position` as well as its sign during controller tuning. If the value is too low, the robot will still tend to drift away, and if it is too high, this term will make the reaction to the control sluggish. If it is significantly too high, it can even lead to uncontrollable robot oscillations caused by the signal lag that is introduced by the moving average filter.

The part of the controller that controls the steering adjusts the value of `diff_pwm`, since it needs to be applied to both motors in different directions to generate the differential steering effect. To make the robot react to the `steer` input of the RC controller, we use a line that is very similar to the one handling the `throttle` input:

```
// Convert steer input to PWM value.
diff_pwm += kp_steer * (float)(steer::pulse_width_us -
kPwmCenter);
```

And to prevent the robot from unintentionally spinning around its vertical axis, we add a `D-control` term for the steering action. Fortunately for us, we can directly measure the rate at which the robot rotates with the gyroscope, so this term can be programmed as follows:

```
diff_pwm += kd_rotation * gyro.gyro.y;
```

As with the other gains, we need to find the values for `kp_steer` and `kd_rotation` during controller tuning.

These are all the control terms that together produce the motor PWM commands, and we are almost done with the software part! There are just a couple more tweaks that we still need to make to the control system to improve its performance and make the robot easier to operate. Let us look at them!

## Motor deadband compensation

You may have noticed already when you were experimenting with the Bluetooth-controlled line follower from the previous chapter that it takes a certain PWM value to get the motors started. A PWM value of `1` will certainly not make them spin; the minimum value is more likely in the range between 20 to 60. This is mostly due to friction in the motor and gearbox. Let us call the range of PWM values that do not lead to any motor movement the **deadband**, a common term in control theory. This deadband is a problem for our self-balancing robot, as we need smooth control around the balanced position. The deadband characteristic, however, means that the robot must pitch quite a bit before the controller output gets large enough for the wheels to start moving, leading to a jerky rocking rather than graceful balancing. A straightforward way to get around this problem is to compensate for the deadband with a helper function that nonlinearly maps an input PWM value to an output value, as shown in *Figure 13.10*:

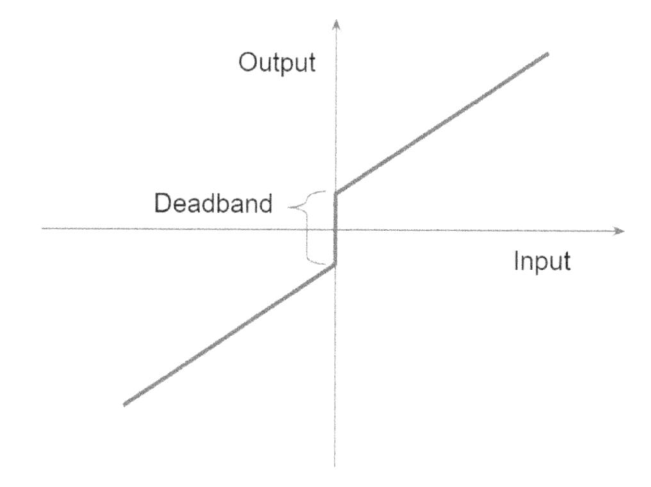


Figure 13.10 – Deadband compensation applies a constant offset with the same sign as the input

Whenever the desired input PWM is positive, our deadband compensator adds a constant offset, and when it is negative, the deadband compensator subtracts the same offset. The following function implements this compensation:

```
int deadbandCompensation(int input, int deadband_offset) {
  if (input > 0) {
```

```
        return input + deadband_offset;
    } else {
        return input - deadband_offset;
    }
}
```

We simply need to pass the output of our controllers to it and use the compensator's output to drive the motors:

```
left::motor.setPwm(deadbandCompensation(pwm + diff_pwm,
kDeadbandPwm));
right::motor.setPwm(deadbandCompensation(pwm - diff_pwm,
kDeadbandPwm));
```

It is good to think twice before adding nonlinear terms like this one to the control system, as they make tuning and debugging even more complicated. However, this function can help us improve the balancing of our robot quite a bit. Nonetheless, you should start with small values for `kDeadbandPwm` and try to keep it smaller rather than larger.

## Auto on/off

When you turn on the robot, it should be calmly lying on the floor to get good readings for the gyro calibration. And we do not want the wheels to spin as soon as it is done calibrating, since there is no hope that the robot could get up on its own. Instead, we need to manually put the robot in its upright position, and the balance control should automatically start controlling the motors only when it is upright. To do that, we introduce a global flag, `motor_enabled`. Only when this flag is set to `true` will we let the controller send non-zero PMW values to the motors. It is initialized to `false` and set to `true` inside the control task when the robot is close to upright:

```
if ((abs(kPitchOffset - pitchFilter.get_angle_filtered()) <
0.01)) {
    motors_enabled = true;
}
```

Similarly, we want the robot to turn off the motors when it has fallen, or when we pick it up and lay it on the ground. To achieve this, we add an `if` statement to the control task that checks whether the robot tilt exceeds a threshold from which it cannot recover, in which case, we set the `motors_enabled` flag to `false` again:

```
if (abs(rad2deg(pitchFilter.get_angle_filtered())) >
kCutoffPitchDeg) {
  motors_enabled = false;
}
```

Since we define kCutoffPitchDeg in degrees (because it is more intuitive for most people) but the pitchFilter operates in radians, we need to convert the pitch filter output to degrees first. Our very own rad2deg() helper function does this for us:

```
float rad2deg(float rad) {
  return 180.0 * rad / PI;
}
```

PI is a constant that the Arduino IDE automatically provides.

## Parameter tuning

As you have seen, there are quite a few parameters in this code that need tuning, notably the feedback filter parameters and controller gains. Controller tuning is the process of changing the parameters, running the robot to see the effect of the change, and repeating the process until the result is good. To change parameters, you can either make the change directly in your program and upload it to your Arduino or use the Serial interface and the Arduino IDE's Serial Terminal (or your phone over Bluetooth) to make changes.

> **Note**
>
> Tuning all the parameters of your robot can be time-consuming. However, it is a great way to gain an intuitive understanding of the underlying algorithms, and the experiments you will do are time well spent to get a feel for working with control systems. Be patient, act methodically, and be careful to not damage your robot in the process.

Controller tuning can easily become a tedious and frustrating experience if you do not follow a few important principles and have an order in mind (or better, on paper) in which you will find the parameter settings.

Most importantly, only ever change one parameter at a time. If you change more than one setting between experiments, it can be close to impossible to understand which effect what change had.

For a wheeled robot like this one, always start testing with the wheels in the air after you change a control parameter. This allows you to verify that the parameter has the correct sign and at least the right order of magnitude before you put the robot on the ground and risk potential damage due to it falling over or running into things.

First, you need to tune the filter parameters that affect the feedback path of your controllers. These are the window size and update rate of the velocity filter in the QuadEncoder library, and the filter coefficient of the complementary pitch filter. You can print the wheel velocities and the estimated pitch angle, respectively, to visualize them in the Serial Plotter as you tune the parameters. You will want to find parameters for both filters that give you responsive outputs with minimal noise. Once the feedback parameters are tuned, you can move on to the controller gains.

To tune the controller gains, you want to start with all of them set to zero. Then, increase `kp_balance` until the robot shows the right reaction (wheels turning forward when the robot leans forward, and vice versa) and adjust it until the robot achieves a stable albeit wobbly balance. Then, do the same for `kd_balance` to make the balancing steadier, increasing its magnitude until it starts to cause oscillations.

> **Note**
>
> The length of your phone mount and the mass of your phone will affect the right value of these two parameters. It is necessary to have the phone mount extended and a dummy mass that represents your phone attached to it to find the right balance gains. Do not use your actual phone at this stage, as there is a good chance that it would get damaged in the process.

Do the same procedure for `kd_position` and `kd_rotation`. If your robot is like the example in this chapter, the values in the code on GitHub should be in the right ballpark for your robot and thus a good starting point for your tuning procedure.

Lastly, you need to find values for `kp_steer` and `kp_throttle` that give you the desired reaction to the control inputs.

## Testing and practice

With the controller parameters dialed in, your robot should be fun to drive around just like a radio-controlled car. It will take a little while to get used to it and you might want to fine-tune some parameters, but eventually, you will get good at driving it around precisely. You will notice that it tends to drift away from its current position and does not naturally stand in one place. You need to constantly control it to hold its position. That is because there is no term in our controller that makes use of the wheel position. It is possible to improve on this, but these improvements are not included in the example code for clarity – it has been a long chapter already!

# Telepresence

Once you have gained confidence in the robot and your ability to steer it around crash-free, it is time to try out the telepresence functionality. You will need to attach your phone to the robot's phone holder for this. It is a good idea to add a good amount of padding around your phone to protect it from impact, should your robot fall for any reason.

You can use any video conferencing application such as Google Meet, Zoom, or Skype to call your phone from your laptop. Some applications even allow you to connect your phone and laptop with only one account. Once the connection is established and you see your phone's video feed on your laptop screen, start the robot and drive it around in **first-person view** (**FPV**) mode. Have fun, and don't crash!

## Summary

In this chapter, we have built an advanced self-balancing robot, based on the line follower of the previous chapter. We have learned many new control techniques and concepts that you can use to design and build more advanced control systems. By tuning all the parameters of your robot, you have gained an intuitive understanding of the filters and controllers we implemented in this chapter, which will prove very useful for future projects. There are many ways to keep learning. You can dive deeper into control systems design, make your robot sturdier, add an automatic kick-stand, add more remote functionalities (why not add a simple arm?), implement an automatic position-hold mode, and much more.

In the next chapter, we will wrap up with a look at what we have learned throughout this book, what else there is to learn, a birds-eye view of the entire field of robotics, and ways to start your career in robotics based on what you have learned in this book.

## Further reading

Self-balancing robots are a popular advanced robotics project, so you can find a lot of interesting information about related projects on the internet. Here are a few examples:

- A short blog with a few more plots showing the complementary filter's performance: `https://www.electronicsforu.com/electronics-projects/prototypes/want-build-self-balancing-robot`.
- Another little self-balancing robot using an IMU and complementary filter: `https://www.instructables.com/Arduino-Self-Balancing-Robot-1/`.
- An interesting self-balancing robot project using stepper motors, with accompanying videos. The video on parameter tuning is especially interesting: `https://www.instructables.com/HeadBot-a-Self-Balancing-Robot-for-STEM-Learning-a/`.
- A very fast radio-controlled self-balancing robot using a high-power BLDC direct drivetrain: `https://www.youtube.com/watch?v=k_vYA-xRXyo`.
- You can control a robot at the other end of the world using the key press tones of your video call, rather than a remote control. The underlying technique is called **Dual-Tone Multi-Frequency (DTMF)** decoding. You can see an example here: `https://www.instructables.com/Using-MT8870-DTMF-Decoder-With-Arduino/`.

# 14
# Wrapping Up, Next Steps, and a Look Ahead

In this chapter, we will take a step back and look at what we have learned throughout this book, and how it fits into the bigger picture of robotics. We will explore what you can study to develop your robotic skills even further, depending on what excites you the most. We will venture a look ahead based on current trends in robotics to give you some inspiration on how to get involved and make an impact.

The chapter is structured as follows:

- A quick recap of what we have covered
- The bigger picture
- Ideas and guidance for your next steps in robotics
- Current trends in robotics research and development
- How to start your career in robotics

## Technical requirements

There are no technical requirements for this chapter. You can sit back, relax, and enjoy the chapter.

## A quick recap

Having learned as much as you did throughout this book, it is helpful to take a moment to recap what you have learned and how it all fits together. So let us take a quick look back.

We learned how to build robots with Arduino boards and other off-the-shelf components. Creating robots requires skills from three different engineering disciplines: mechanical engineering, electrical engineering, and computer science (or programming).

We started by getting to know the Arduino platform. Arduino boards let us run small robot programs and interface with a large variety of hardware components. We covered writing programs for Arduino and understanding what Arduino board to select based on CPU specifications (clock rate and register size), memory size (FLASH, SRAM, and EEPROM), and available interfaces (SPI, I2C, serial, and so on). These topics are all part of computer engineering.

We also learned how to use the Arduino to interface with a variety of components that are useful for building DIY robots. This included controlling different motors (RC servos, DC, stepper, and BLDC motors), reading sensors over various interfaces (digital and analog), controlling LCDs, controlling multicolor LEDs, and generating sound output. These topics lie at the intersection of computer engineering and electrical engineering. Here, we not only need to write the program but we also need to understand how to connect our hardware and take care that voltage levels match, that wires are appropriately sized for the expected currents, and that physical signals do not interfere with each other. We also learned how to select batteries, sense voltages, and currents, and how to design a basic power system to safely and efficiently power your robot. These are topics that squarely lie in the electrical engineering domain.

Lastly, we learned how to integrate all of these components into an actual robot. This requires some basic mechanical engineering. We saw that we can build robots from scratch with simple tools and basic materials such as wood, screws, bolts, double-sided tape, and cable ties. And we also saw that using 3D **Computer-Aided Design** (**CAD**) combined with a small and inexpensive desktop 3D printer even makes building the complex mechanical features of our robots really easy.

All of these lessons have given you a solid foundation to start your own journey as a DIY robot builder. If you have a robot in mind that you want to build, now you know what components are available to you, how to select them, integrate them, and build and program your robot to make your dream a reality.

To exemplify the process, we followed the more theoretical chapters with two example projects that show a lot of this learning in action. First, we built an advanced version of the classic line-following robot, with two motors, an Arduino Mega, a PixyCam camera sensor, and a Bluetooth interface chip, which lets you control the robot from your phone. We methodically selected the components, wrote a modular program that lets us test components individually, and implemented a closed-loop controller that lets the robot follow a line on the ground completely autonomously. We barely scratched the surface of what you can do with this robot; we just covered enough to get you started trying out all the ideas you might have in mind (for example, going much faster, handling intersections, and maybe even going backward). Thanks to the Bluetooth link of this robot, you can also use it to get started with app development and create your own, more capable remote control phone app. Or you can add one of the other wireless interfaces we covered (Zigbee and Wi-Fi) and explore (or write your own) ground station software for your PC to control your robot and view live telemetry.

And finally, we modified the line-following robot and built an impressive, remote-controlled, self-balancing telepresence robot. This project allowed us to explore a few new topics that are important for creating advanced Arduino robots. We saw how to change the microcontroller's settings *under the hood* to increase the PWM frequency for motor control. And we learned about two methods for sensor signal processing that allowed us to estimate quantities that we cannot directly measure. A moving average filter helped us to estimate the wheel velocity from the motor encoder position data, and a complementary filter enabled our robot to estimate its pitch angle from the IMU's gyroscope and accelerometers. This is an advanced Arduino robot in many regards, and it has tremendous potential for further improvements that you can explore. You have probably spent quite a bit of time testing and tuning parameters and, in the process, you gained experience and intuition about the behavior of control systems. Better intuition is the superpower that will allow you to build new robots faster. It comes from experience, and experience comes from spending time on hands-on work, making real-world robots work. Reading encoders, decoding PWM, and sampling the IMU at high frequency is getting close to what the humble 8-bit, 16 MHz Arduino Mega can handle. To unlock more capabilities such as even better control, more complex pitch estimation, PPM decoding to read all RC channels, and additional hardware, this robot is a good candidate for experimenting with a more capable board such as the Teensy 3.6.

# The bigger picture

The field of robotics is vast, so let us look from a birds-eye perspective to see where our learning from this book fits in the bigger picture.

Robotics not only encompasses several engineering disciplines and many of their specialized subfields but it also covers a myriad of applications. We touched on several disciplines, but we focused on the application of small, wheeled mobile robots. Some of the other important areas for applications of robots include the following:

- Stationary robot arms, used in manufacturing and material handling
- Smart conveyor systems for automatic material handling in warehouses and process logistics
- Flying robots for autonomous delivery, aerial photography, and surveillance
- Space robots for the exploration of harsh, extraterrestrial environments
- Large mobile robots, such as automated equipment for agriculture or mining
- Social robots for companionship and everyday support
- Humanoid robots for advanced research applications

Everything we discussed in this book applies to these classes of robots, and you have a good grasp on many of the important basics for building any of these systems now. But building advanced, product-level robots for most of these fields requires you to continue to learn and gain experience.

Most importantly, we need to keep in mind the limitations of the Arduino platform. We chose it for its great accessibility and the power it gives us to create entire robot systems from scratch easily and cost-effectively. But the computational capabilities of what even the most powerful Arduino board can do are not enough for many modern, real-world robot applications. Robots such as self-driving cars need a lot more brain power. In most advanced robot systems, there is an entire **stack**

**of technologies** to realize all the required capabilities. The stack ranges from hardware interfaces and real-time systems (low-level) up to complex perception methods, planning algorithms, and database interfaces (high-level). Embedded systems such as our Arduino robots can be part of the low-level components of this stack – for example, for motor control or safety-critical monitoring functions. The absence of an operating system makes embedded systems great for real-time control tasks and safety systems where the utmost reliability is required. But the embedded systems are just parts of the entire robot, and they communicate with the higher-level components.

On the high-level side of a robot's technology stack, you will often find one or more powerful computers that more closely resemble (or even exceed) your laptop or workstation in terms of performance, rather than an Arduino microcontroller. These computers typically run a full-fledged Linux-based operating system that is easier to use but is less well suited for real-time control. Modern robot systems often further contain **Graphics Processing Units** (**GPUs**) or other hardware accelerators for image processing and specialized **Machine Learning** (**ML**) applications.

The big takeaway is that robotics is a multi-faceted field and modern robots are increasingly complex systems, both in terms of hardware and software. These full-stack robots are developed by entire teams of engineers that specialize in various aspects of the technology stack, from designing custom electronics to optimizing controllers, calibrating sensors, or optimizing motion planning algorithms.

It makes sense to start your journey into the field of robotics at the low-level end of the robotics technology stack, as you did, by creating your own Arduino robots. This is the fastest way to gather invaluable hands-on experience and find out what aspects of robotics excite you the most to direct your further studies. Depending on in which direction you want to grow, your next projects and areas of study may vary. Let us discuss a few of your options.

# Your next steps

Of the many paths you could take from here, this section will list a few suggestions:

- Master DIY robotics with Arduino and beyond
- Focus on the mechanical engineering of robots
- Focus on the electrical engineering of robots
- Focus on the computer engineering and computer science aspects of robotics
- Focus on data-driven methods for robotics ML

## Becoming a DIY robotics master

If you want to continue creating ever more advanced Arduino robots and mastering DIY robotics, you already have a good sense of what skills you need to hone: everything that we covered or touched upon in this book.

For building robots, it quickly pays off to pick a (free) CAD software (such as Onshape) and aim to get increasingly proficient at it. Practicing CAD is easy; you can simply pick any object that you see and try to create a 3D model of it in CAD. Get comfortable creating various geometries from scratch, designing parts in the context of other parts, and combining multiple parts into assemblies. Just like we aim to write code that can be easily modified with only a few changes, well-made, parametric CAD models can be easily modified to accommodate changes in your design, such as using larger motors or bigger batteries.

Along with CAD, aim at getting access to a 3D printer (if you do not already have one) and practice using it. Depending on your printer and slicer software, there can be many settings to figure out, different filaments to try, and calibration procedures to run. Investing the time in establishing a smooth 3D printing toolchain (from CAD model to physical part) pays off fast when building robots. It will save you a lot of time and money if you can simply print custom-designed parts instead of having to make them some other way. 3D printing also enables designing parts that you could not make any other way, simplifying your robot design, and allowing for more elegant design solutions. It takes a lot of practice to be able to efficiently design parts that just work the first time. So, try to design and 3D-print a lot of parts. Even parts outside your robotics projects will help you gain experience – for example, phone holders, door stoppers, fidget toys, or anything that fits on your printer.

On the electrical side, the skill to master is soldering. Good soldering skills are essential for building robots, and creating reliable, high-quality solder joints is harder than it looks. It is all too common in the world of DIY robotics that a bad solder joint is the reason for a robot malfunctioning, leading to long and frustrating debugging sessions or even robot damage. If you can, use a high-quality, well-maintained soldering iron (or better yet, a soldering station) and keep your tips in good shape by regularly cleaning them. Practice soldering on wires of various diameters and materials – soldering thin wires with PVC insulation requires different temperature settings and soldering techniques than soldering thick wires with silicone insulation. Practice soldering pin headers to breakout boards and learn how to inspect, desolder, and rework solder joints.

Learn more about the various electrical components you might need to build your robots (ICs, LEDs, diodes, capacitors, resistors, sensors, and so on) and practice reading schematics and component datasheets. Datasheets tend to be lengthy documents, and it takes practice to be able to quickly extract the information that you need to decide whether the component is the right choice for your project. Once you know how to effectively read datasheets, they become an invaluable source of information for your circuit design process. They often contain reference schematics for certain applications that can be your fast lane to a working prototype.

Practice drawing electrical schematics to help structure your design process and document your circuits. Drawing circuit diagrams by hand is a great start, but consider moving to a digital tool as soon as possible. For beginners, many **Electronic CAD (ECAD)** or **Electronic Design Automation (EDA)** tools can be confusing and overwhelming, and the barrier to entry is unfortunately quite high. An exception to this is Fritzing, which is aimed to lower this barrier to allow hobbyists to use ECAD (`https://fritzing.org/`). Most of the schematic drawings throughout this book were created with Fritzing. Fritzing is an open source software that you can build from the freely available source code on GitHub, or you can pay to directly download the application for your operating system. Vendors of electronic components targeted toward the DIY community, such as Adafruit and Sparkfun, provide Fritzing libraries for their products via GitHub so that you can easily integrate them into your circuit designs (`https://github.com/adafruit/Fritzing-Library` and `https://github.com/sparkfun/Fritzing_Parts`).

When it comes to programming, getting proficient at writing and reading C and C++ code is what will make you a master Arduino programmer. A practical way to get started is by reading high-quality Arduino libraries on GitHub (for example, the ones published by Adafruit) and making it your goal to understand every detail (`https://github.com/adafruit`). If there is anything that is not clear to you, use resources such as Stack Overflow or C++ textbooks to learn about it (`https://stackoverflow.com/questions`).

An effective way to solidify your understanding is to re-implement interesting code examples from scratch or to explain exactly how the code works to a friend, the wall of your room, or any other object. This might sound silly at first, but explaining code to an inanimate object such as a rubber duck (*rubber ducking*) is an established method of testing your understanding of an algorithm or finding bugs in a program (`https://en.wikipedia.org/wiki/Rubber_duck_debugging`). The other essential tools you will need to be most effective are Git and GitHub. As we saw, almost all relevant software, from Arduino libraries to third-party software such as Fritzing, are hosted on GitHub. Work through the GitHub tutorials to get set up and comfortable using GitHub to get code, store code, and even contribute to open source projects through pull requests (`https://docs.github.com/en/get-started/quickstart/hello-world`). Even without GitHub, using Git locally to manage your program development workflow and benefit from its strong version control features is something you will not want to miss once you get the hang of it. Unfortunately, Git was developed with experienced software engineers in mind, and it can be a little bit intimidating to beginners. However, it is an incredibly useful tool and well worth the time it takes to learn to use it.

Investing time in these areas will be your most effective path toward becoming a master DIY roboticist. There is no substitute for doing a lot of little projects, taking every opportunity to learn something new, and gaining experience along the way. Once all these tools become second nature, you will be amazed at how quickly you can create impressive new robots.

If you want to take the next step from Arduino-based DIY robots to working on a more complex full-stack robotic system, you will need to pick an area to specialize in. Creating increasingly advanced Arduino robots can give you a good idea of the area that fascinates you the most. This is the area you should focus on.

## Focusing on mechanical engineering

If what you want to specialize in is the mechanical design of robots, this section is for you. An effective way to start is by finding ways to improve existing designs. You can think of better ways to design your line-following robot, for example. Try to make it sturdier, more modular, easier to assemble and disassemble, and give it a body. You can go through the same exercise for the telepresence robot.

For more advanced robots, you will need to learn about materials and fabrication methods that we have not discussed in this book. Particularly popular choices for building advanced robots are fiber-reinforced materials, aluminum, and steel. Learn how aluminum and steel can be processed and machined, joined, formed, and bent. Mastering designs for sheet metal fabrication is an especially useful technique for cost-effective designs, and it is supported by modern CAD software packages. You should also learn about other additive manufacturing processes besides fused filament fabrication. **Selective Laser Sintering** (**SLS**) and photopolymer liquid resin 3D printing are particularly useful processes. Unfortunately, getting hands-on experience with these materials and processes is much harder and much more expensive due to the cost of the materials themselves and of the tools and machines. There are plenty of services that let you submit your digital design files (exported from CAD) and that will manufacture them for you from the material and with the process of your choice, such as Protolabs (`https://www.protolabs.com/`) or Xometry (`https://www.xometry.com/`). Even though you do not need access to the machines with this option, this route is still much more expensive than simple PLA printing at home. Therefore, you should spend some time studying the theoretical aspects of designing with these advanced materials rather than purely learning by doing, as we can in the realm of small, homemade Arduino robots. Modern CAD software lets you run a structural analysis known as **Finite Element Analysis** (**FEA**), which can help you to be confident that your parts will perform well, even before you have them manufactured.

When designing mechanical parts for advanced robots, you need to understand all the physical requirements for the part to select the right material and design. The mechanical properties of the part, its strength, and its stiffness are typically the most important, followed by its weight, thermal and electrical properties, material cost, and manufacturability. To understand the mechanical requirements and design parts accordingly, you will want to get a good grasp on **engineering mechanics**, the branch of physics that describes mechanical loads (forces and moments) and their effects on structural materials. Engineering mechanics can be divided into two branches: **statics** describes mechanical systems at rest, and **dynamics** describes the more general case of mechanical systems that move, where additional loads caused by accelerations need to be accounted for.

Lastly, try to get to know as many mechanical components as possible. There is a tremendous variety of standard components available, such as ball bearings, springs, and rubber dampers, that are waiting to be discovered by you. The more components you know about, the easier and faster you can design mechanisms and prototypes. A good way to get a glimpse into the world of mechanical components is the informative website of McMaster Carr (`https://www.mcmaster.com/`).

## Focusing on electrical engineering

If the electrical components and the wiring of your Arduino robots are what fascinate you most, you might want to focus on electrical engineering aspects on your path toward working on more advanced robot systems. Start by making sure that you understand every bit of circuitry in your Arduino robots. Find the schematics of your Arduino boards and all the breakout boards you have, learn how to read them, and understand which physical components on the board match which symbol in the schematics. Use a **Digital Multimeter** (**DMM**) to verify connectivity, voltages, resistances, and capacitance values. Probing various locations of your boards will teach you what you can and cannot measure, and you will certainly see unexpected measurements. Trying to understand all of them will help guide you in your study of basic circuits. Besides classical textbooks, online courses and community sites such as Electronics StackExchange can be tremendous resources (`https://electronics.stackexchange.com/`). There are many advanced open source electronics projects out there, which means that you can study their schematics and learn a lot that way. Other than that, classical textbooks on power electronics, digital, analog, and mixed-signal design, as well as similarly themed online courses, can be invaluable resources. There is a lot to learn when it comes to advanced electronics and circuit design, and you will need to come prepared for a lot of specialized notations and some advanced mathematical concepts.

Next, you will need to learn how to use a capable ECAD tool to design your own **Printed Circuit Boards** (**PCBs**). These tools allow you to do the complete PCB design, from schematic to the full board layout and even specialized production outputs (such as Gerber files). The list of high-quality ECAD software packages includes names such as Altium Designer, Eagle, and KiCAD. KiCAD stands out as being fully open source and completely free to use (`https://www.kicad.org/`). Getting started with ECAD software can be a bit tricky, so you will need to budget some time and follow available tutorials to help you get started. Once you are all set, design your first PCBs. You can design your own Arduino UNO clone based on the UNO's open source schematic, or a shield for your Arduino that has all your robot's components integrated. To make your own PCB a reality, you have two main options. You can either order the PCBs from a fabrication service and populate them yourself (for example, by converting a small oven into a reflow soldering station), or you can go down the more costly route of ordering fully assembled boards (also known as turnkey manufacturing). For bare PCBs, services such as PCBWay or OSH Park are among your options (`https://www.pcbway.com/` and `https://oshpark.com/`). While you can do PCB manufacturing yourself with highly specialized chemical processes or specialized milling tools, it is very rarely worth the investment of time and money if you have other options. Easy-to-use turnkey PCB providers include CircuitHub and Macrofab, among many others (`https://circuithub.com/` and `https://macrofab.com/`). Once you find the tools and processes that let you go from an idea to a physical board and that work well for you, aim to design and test ever more advanced PCBs to hone your craft.

## Focusing on computer engineering and computer science

You may have realized that the hardware of your robot, mechanical or electrical, is not what excites you the most. It is the software that you want to dive deeper into. Consider yourself lucky! To become an expert programmer, you do not need to worry about the cost and time it takes to manufacture physical parts; all you need is the computer that you likely already have. And you have the benefit of trying things out fast, without the fear of breaking anything that takes time and money to repair or remake. Software engineering is an area that lends itself exceptionally well to self-study, with countless books, online courses, and tutorials available to you.

At the beginning of your journey, you need to pick one or two programming languages to focus on. By programming Arduino robots, you have already started to use C++. C++ is also widely used in advanced, real-world robot projects, so you can simply stay with this choice. However, despite its many strengths, such as advanced compilers, type safety, and high performance, C++ can be cumbersome and difficult to master. Another excellent choice at the beginning of your journey is the Python programming language. Python is very user-friendly, readable, and quick to develop in. For these reasons, Python is widely used in robotics research (`https://www.python.org/`).

Once you can write your own computer programs in C++ and/or Python, a great next step is to learn to use the **Robot Operating System** (**ROS**) (`https://www.ros.org/`). ROS is widely used in DIY, research, and even some commercial robotic systems. It acts as the backbone (middleware) of larger robotics software projects, where many individual programs (*nodes*, in ROS terminology) need to communicate with one another, even across different robots. There are many great tutorials, example projects, books, and online courses to help you understand ROS and when and how to use it.

To practice your programming skills and learn from more experienced programmers, get involved with open source robotics projects (such as ROS itself), read lots of code, and start contributing your own features or bug fixes through *pull requests*. Working on an actual project as part of the community and getting feedback from others is a highly effective way of gaining experience quickly.

There are many distinct aspects when it comes to robot software, including topics as diverse as real-time control, behavior generation, computer vision, and motion planning. Eventually, you will find the area that suits your interests best and you can then focus your learning there.

## Focusing on data-driven methods ML

Among the many aspects of computer science and engineering in robotics, few have received as much attention and have risen in relevance as quickly as data-driven methods, summarized as **Machine Learning** (**ML**), sometimes synonymously called **Artificial Intelligence** (**AI**). Historically, programmers wrote detailed robot programs that meticulously executed an exactly defined algorithm. The examples in this book fall into this category. In data-driven methods, the programmer does not need to specify the program's function precisely. Instead, the program learns the details itself, based on data or experience. The classic example is image recognition. It is extremely hard to write a program that reliably recognizes a cat (or any other object category) in all images of cats on the internet. The variety of images is just too large. It turns out to be much easier to collect a (large) set of images of cats and use these to let an image classifier learn for itself how to reliably recognize cats. The applications of ML techniques range far beyond image classification and play a key role in many parts of modern robotic systems.

The field of ML has rapidly evolved in the past decade, and there are many textbooks, research papers, established methods, and datasets available for you to dive right into. Once you understand the basic concepts, such as classification, regression, supervised and unsupervised learning, as well as core algorithms from linear regression to deep learning, you can start experimenting with one of the numerous toolboxes and frameworks. Many open source ML applications and research projects use Python, and PyTorch is a great tool to start your hands-on learning by implementing your own ideas or re-implementing interesting approaches (`https://pytorch.org/`).

# Trends in robotics

There are more than enough highly active areas in robotics today to keep all of us busy for a while. Let us take a quick look at some of the current trends and areas that might become even more important soon (in no particular order).

## ML/AI

Data-driven methods (ML or AI) have transformed robot perception and made their way into robot control, and impressive progress is still being made by researchers all over the world. With ever more powerful **Large Language Models** (**LLMs**), such as OpenAI's GPT-3, and their astonishingly broad-ranging capabilities, it seems likely that ML will continue to have a significant impact on the field of robotics. We may soon see robots that are more autonomous and can react smarter and more independently to unforeseen and challenging situations than is currently possible, thanks to advances in this area.

## Cloud robotics

Most robots to date are independent actors. The idea behind cloud robotics is to connect robots over the internet and give all of them access to shared databases to share their experiences and learn from one another, and to allow centralized control and robot fleet management. In some sense, you can think of a cloud robotic system as one robot that has many small, locally distributed robots as its body parts, or, put the other way around, a lot of small robots with a shared brain. There are still many engineering and research challenges regarding communications and control to overcome before massive cloud robotic systems become widespread, but we may start to see increasingly more of them soon.

## Human/robot collaboration

In many applications today, such as industrial manufacturing, robots work spatially separated from humans to guarantee safety. Robots are often powerful enough to cause harm and damage and lack the spatial awareness, the understanding of human behavior, and the required sensing to reliably avoid accidents when they share a workspace with humans. This is a major restriction that holds robotics back from many useful applications, such as in healthcare, customer service, or small-scale manufacturing. With better hardware, more capable perception systems, and novel control methods, a safe physical collaboration between robots and humans may become much more widespread soon and is a trend to keep on your radar.

## Soft robotics

When you think of the mechanical structure of robots, you are probably imagining something rigid. Most robots today are, in fact, rigid – made to be as stiff as possible to move as accurately and predictably as possible. There are downsides to this paradigm, for example, when it comes to working with humans or adapting to damage. Inspired in part by the softness of most natural organisms, the field of soft robotics aims to develop robots that are soft, flexible, and inherently safer to be around than traditional stiff robots. There are many exciting engineering and control challenges waiting to be solved to make fully soft robots, but soft robotic grippers are gaining popularity already.

## Robots as consumer products

Currently, most robotics applications are in commercial settings where robots are programmed, used, and maintained by professional experts. Despite significant efforts by big companies, there are hardly any robots directly sold to and used by consumers. Floor-cleaning robots are the one big exception. With substantial research and development efforts currently underway, and the promise of a giant market for consumer robots, it may only be a matter of time before we will see advanced robots that are easy to use, useful, cheap, and reliable enough to enter the consumer market.

## Robots for autonomous data acquisition

The robotics research community has developed an impressive range of robot types that can traverse almost any terrain with treads or legs, swim, dive, or fly. Often developed with disaster response applications in mind, some of these technologies have found applications as autonomous data collection machines. There are now legged robots roaming and inspecting remote oil platforms, legged and flying robots monitoring progress on construction sites, and flying robots regularly surveying critical infrastructure such as bridges. With the decreasing cost and increasing availability of such solutions, we might see ever more autonomous data collection robots working in remote locations and hard-to-access areas.

## Robots in logistics

With the rise of e-commerce and the resulting rise in large-scale warehouses for consumer goods, the need for more efficient ways to operate them is ever-increasing. Modern warehouses are already highly automated, and robots such as autonomous vehicles and stationary and mobile manipulators play a key role in this trend. Companies such as Amazon and Ocado Technology are developing advanced cloud robotics solutions to create a new generation of flexible and efficient warehouses. In addition, there are many companies and start-ups working on autonomous cargo trucks – essentially, giant robots that can transport goods between logistic centers.

## Humanoids as multi-purpose robots

While it may still be far away, academic and industry research efforts are underway to develop multi-purpose robots that are not optimized for just one task (such as moving boxes in a warehouse) but are truly multi-purpose. Much of the inspiration for this comes from the human body and mind, which enable humans to perform many tasks in a wide variety of circumstances. For that reason, a robot modeled after the human body, a so-called humanoid robot, is often considered the most useful, multi-purpose robot morphology. Humanoids are, therefore, at the center of several advanced robotics research projects.

# How to start your career in robotics

If you are considering robotics not just as a hobby but as a career path, you might be wondering how to get started. The most straightforward path starts with a college education in mechanical engineering, electrical engineering, or computer science. During this time, consider getting involved with robotics clubs, contributing to open source robotics projects, and securing internships at robotics companies or in related industries. This will put you in a good position to apply for your dream job after graduation, or help you build the network you need to get your own start-up company off the ground. If you are looking for a career in robotics research, consider joining a Ph.D. program.

If you do not have a formal engineering education, you can still learn all the required skills from books, tutorials, and online courses. It will require considerable time and effort, however. Among the many materials freely available to you, the robotics courses from MIT OpenCourseWare are especially noteworthy (`https://ocw.mit.edu/search/?q=robotics`). Try to connect with like-minded people, work on as many projects as you can, compete in robotics competitions, and contribute to or start your own open source robotics projects. This will help you to gain solid expertise and add credibility to your resume, making you a competitive candidate when you apply for robotics jobs.

If you are already an engineering professional working in a different field and want to switch to robotics, the switch might be easy, depending on your desired role. Try to find a robotics position that resembles your current role as closely as possible. Once you successfully make the move into the robotics industry, you can gain experience, grow your skill set and professional network, and shape your career path in the direction you want.

# Summary

In this chapter, we looked back at all the previous chapters and recapitulated what we have learned, and how it fits into the bigger picture of robotics. We discussed the possible next steps of your educational journey into the world of robotics, tools to learn, and projects to work on, depending on what excites you the most. We also talked about several current trends in robotics that may become even more prominent topics in the near future to give you an idea of where the field of robotics is headed. And lastly, we briefly outlined how you can turn robotics from an interest into a career, starting where you are right now.

If you have made it all the way here, you hopefully learned a lot about DIY robotics with Arduino and important aspects of robotics in general. Thank you so much for reading this book. Have fun with your new skills, keep building, making, and learning, and make your dream robot projects come true! And if you liked the book, please consider recommending it to anyone you think might enjoy it, too.

# Index

## Symbols



## A

# B

# C



# D

# E

## K



## L



## M

# N



# O



# P

# R

## S

# T

## U

## Z

packtpub.com

Subscribe to our online digital library for full access to over 7,000 books and videos, as well as industry leading tools to help you plan your personal development and advance your career. For more information, please visit our website.

## Why subscribe?

- Spend less time learning and more time coding with practical eBooks and Videos from over 4,000 industry professionals
- Improve your learning with Skill Plans built especially for you
- Get a free eBook or video every month
- Fully searchable for easy access to vital information
- Copy and paste, print, and bookmark content

Did you know that Packt offers eBook versions of every book published, with PDF and ePub files available? You can upgrade to the eBook version at packtpub.com and as a print book customer, you are entitled to a discount on the eBook copy. Get in touch with us at customercare@packtpub.com for more details.

At www.packtpub.com, you can also read a collection of free technical articles, sign up for a range of free newsletters, and receive exclusive discounts and offers on Packt books and eBooks.

# Other Books You May Enjoy

If you enjoyed this book, you may be interested in these other books by Packt:

**Learn Robotics Programming - Second Edition**

Danny Staple

ISBN: 978-1-83921-880-4

- Leverage the features of the Raspberry Pi OS
- Discover how to configure a Raspberry Pi to build an AI-enabled robot
- Interface motors and sensors with a Raspberry Pi
- Code your robot to develop engaging and intelligent robot behavior
- Explore AI behavior such as speech recognition and visual processing
- Find out how you can control AI robots with a mobile phone over Wi-Fi
- Understand how to choose the right parts and assemble your robot

**Raspberry Pi Pico DIY Workshop**

Sai Yamanoor, Srihari Yamanoor

ISBN: 978-1-80181-481-2

- Understand the RP2040's peripherals and apply them in the real world
- Find out about the programming languages that can be used to program the RP2040
- Delve into the applications of serial interfaces available on the Pico
- Discover add-on hardware available for the RP2040
- Explore different development board variants for the Raspberry Pi Pico
- Discover tips and tricks for seamless product development with the Pico

## Packt is searching for authors like you

If you're interested in becoming an author for Packt, please visit `authors.packtpub.com` and apply today. We have worked with thousands of developers and tech professionals, just like you, to help them share their insight with the global tech community. You can make a general application, apply for a specific hot topic that we are recruiting an author for, or submit your own idea.

## Share Your Thoughts

Now you've finished *Practical Arduino Robotics*, we'd love to hear your thoughts! Scan the QR code below to go straight to the Amazon review page for this book and share your feedback or leave a review on the site that you purchased it from.

`https://packt.link/r/1804613177`

Your review is important to us and the tech community and will help us make sure we're delivering excellent quality content.

## Download a free PDF copy of this book

Thanks for purchasing this book!

Do you like to read on the go but are unable to carry your print books everywhere? Is your eBook purchase not compatible with the device of your choice?

Don't worry, now with every Packt book you get a DRM-free PDF version of that book at no cost.

Read anywhere, any place, on any device. Search, copy, and paste code from your favorite technical books directly into your application.

The perks don't stop there, you can get exclusive access to discounts, newsletters, and great free content in your inbox daily

Follow these simple steps to get the benefits:

1. Scan the QR code or visit the link below

`https://packt.link/free-ebook/9781804613177`

2. Submit your proof of purchase
3. That's it! We'll send your free PDF and other benefits to your email directly

Printed in the USA
CPSIA information can be obtained
at www.ICGtesting.com
CBHW081744160724
11648CB00022BB/173

9 781804 613177